D1524387

ACS SYMPOSIUM SERIES **456**

Selective Fluorination in Organic and Bioorganic Chemistry

John T. Welch, EDITOR
State University of New York at Albany

Developed from a symposium sponsored
by the Division of Fluorine Chemistry
at the 199th National Meeting
of the American Chemical Society,
Boston, Massachusetts,
April 22–27, 1990

American Chemical Society, Washington, DC 1991

Library of Congress Cataloging-in-Publication Data

Selective fluorination in organic and bioorganic chemistry / John T. Welch, editor.

p. cm.—(ACS symposium series; 456)

"Developed from a symposium sponsored by the Division of Fluorine Chemistry at the 199th National Meeting of the American Chemical Society, Boston, Massachusetts, April 22–27, 1990."

Includes bibliographical references and indexes.

ISBN 0–8412–1948–6

1. Fluorination—Congresses. 2. Organofluorine compounds—Congresses.

I. Welch, John T. II. American Chemical Society. Division of Fluorine Chemistry. III. American Chemical Society. Meeting (199th: 1990: Boston, Mass.) IV. Series.

QD281.F55S45 1991
547'.02—dc20 90–26633
CIP

The paper used in this publication meets the minimum requirements of American National Standard for Information Sciences—Permanence of Paper for Printed Library Materials, ANSI Z39.48–1984. ♾

PRINTED IN THE UNITED STATES OF AMERICA

ACS Symposium Series

M. Joan Comstock, *Series Editor*

Foreword

THE ACS SYMPOSIUM SERIES was founded in 1974 to provide a medium for publishing symposia quickly in book form. The format of the Series parallels that of the continuing ADVANCES IN CHEMISTRY SERIES except that, in order to save time, the papers are not typeset, but are reproduced as they are submitted by the authors in camera-ready form. Papers are reviewed under the supervision of the editors with the assistance of the Advisory Board and are selected to maintain the integrity of the symposia. Both reviews and reports of research are acceptable, because symposia may embrace both types of presentation. However, verbatim reproductions of previously published papers are not accepted.

Contents

BIOLOGICAL APPLICATIONS

INDEXES

Preface

INTEREST IN ORGANOFLUORINE CHEMISTRY IS GROWING at a rapid pace. Fluorine as a substituent can significantly affect the properties of molecular systems. Its high electronegativity and small atomic volume have been used to probe reactivity—especially that of biologically important molecules.

This book brings together works of authors who are active in theoretical studies of organofluorine compounds, engaged in the development of new preparative methods and exploring the utility of specifically fluorinated compounds in biological systems. The volume is interdisciplinary in coverage and provides a forum for the exchange of information among theoreticians, synthetic chemists, biochemists, medicinal chemists, and bioorganic chemists studying and using fluorine-containing compounds. Readers with specialized interests will find new opportunities in their specific areas, as well as the latest progress in other areas.

In addition to reactivity and theoretical aspects, this volume provides significant coverage of biological applications. The contributions from industry, academia, and government laboratories emphasize the fact that research on organofluorine compounds is far from an ivory tower endeavor and, in fact, that selective fluorination does provide a useful alternative for the study of biological problems.

Acknowledgments

The preparation of this volume would not have been possible without the willing cooperation of both the contributors and the reviewers. I would like to thank Professor H.-J. Hansen of the Organisch-Chemisches-Institut of the Universität Zürich for his hospitality and support during the sabbatical leave where a large part of the editorial obligations associated with this volume were completed. It is also a pleasure to thank Professor D. Seebach for the encouragement he gave me during my sabbatical stay at the Laboratorium für Organische Chemie of the ETH Zürich. The symposium from which this book was derived was organized during the time I spent there. Finally, it would not have been possible for the symposium to have been held at all without the financial support of the Division of Fluorine Chemistry of the American Chemical Society and the Petroleum Research Fund of the American Chemical Society.

JOHN T. WELCH
State University of New York
Albany, NY 12222

November 5, 1990

Chapter 1

The Effects of Selective Fluorination on Reactivity in Organic and Bioorganic Chemistry

John T. Welch

Department of Chemistry, State University of New York, Albany, NY 12222

The influence of fluorine on the physical, chemical and biological properties of molecules is described. A discussion of the effect of fluorination on bond lengths and bond strengths is presented in the course of an analysis of the steric demand of fluorinated substituents. These steric effects as well as electronic interactions are illustrated in a discussion of fluorinated vitamin D_3 and prostaglandin and thromboxane analogs.

The current level of interest in the preparation of selectively fluorinated compounds is indicated by the increasing number of publications and presentations in this area (*1*). It has been known for some time that fluorine can have profound and often unexpected effects on activity. Selective fluorination has been an extremely effective synthetic tool for modifying and probing reactivity. Replacement of hydrogen or hydroxyl by fluorine in a biologically important molecule often yields an analog of that substance with improved selectivity or a modified spectrum of activity (*2-21*). A number of very valuable monographs and other general reviews are available for guiding chemists in the organic chemistry of fluorine (*22-28*).

Historical Perspective

Although hydrogen fluoride was discovered by Scheele in 1771 (*23*), molecular fluorine itself wasn't prepared until 1886 by Moissan (*29*). Organofluorine chemistry really began to unfold with the work of Swarts (*30*) on the preparation of fluorinated materials by metal fluoride promoted halogen-fluoride exchange reactions. The commercial utility of organofluorine compounds as refrigerants, developed by Midgley and Henne (*31*), further accelerated the growth of the field by virtue of the economic incentives involved. Wartime requirements stimulated research on thermally stable and chemically resistant materials which led to an increased interest in highly fluorinated and perfluorinated substances. Many of the exciting developments of this period have been reviewed elsewhere (*32-34*). A synopsis of the growth of organofluorine chemistry could not be complete without recognition of the enormous impact of electrochemical fluorination techniques. The electrolysis of organic compounds in anhydrous hydrogen fluoride was discovered and developed by Simons (*35-39*).

0097–6156/91/0456–0001$06.00/0

It was the pioneering work of Fried on preparation of 9α–fluoro-hydrocortisone acetate (**1**) (*40*) that led to the first significant successful application of selective fluorination for the purpose of modifying biological activity. This publication marked the beginning of a new era when medicinal chemists and biochemists routinely introduced fluorine as a substituent to modify biological activity.

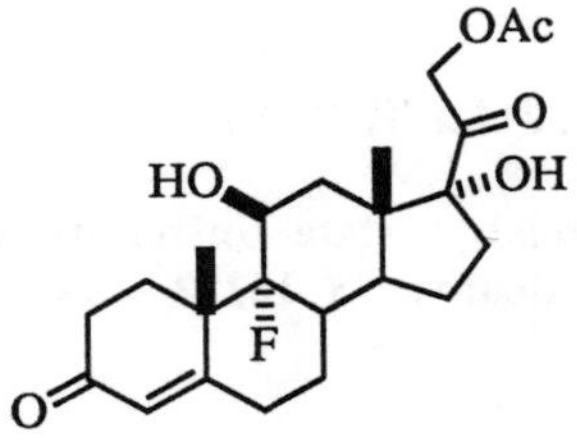

9α-Fluoro-hydrocortisone acetate

1

Structure and Bonding

Fluorine is the most electronegative element with a Pauling electronegativity of 4 as compared to 3.5 for oxygen, 3.0 for chlorine or 2.8 for bromine. It is this property that is the apparent origin of many of the profound differences observed when fluorinated and non-fluorinated molecules are compared. Fluorine forms the strongest bond to carbon of the halogens, e.g., the carbon-fluorine bond energy is 456-486 kJ per mole and the carbon-chlorine bond 350 kJ per mole. For an additional comparison the carbon-hydrogen bond energy varies from 356 to 435 kJ per mole. Important also is the observation that the carbon-fluorine bond is shorter, 1.31 Å, than the other carbon-halogen bonds, for instance, the carbon-chlorine bond at 1.78 Å. Again for comparison, the carbon-hydrogen bond is 1.09 Å long and the carbon-oxygen bond measures 1.43 Å. However the bond lengths vary in an interesting manner in multipli-fluorinated methanes. With increasing fluorination the bond lengths shorten and the bond strengths increase. This phenomena is unique among the halogens (*41*).

Several alternative explanations of this phenomena have been put forth. Most simply, the contraction has been suggested to be a result of the donation of electron density from the non-bonded electron pair on fluorine into an adjacent carbon-fluorine bond (22). Alternative explanations have suggested that bonding with fluorine causes an increased likelihood that the p electrons of carbon will be preferentially shared with fluorine and thus the carbon fluorine bond will have more s-character and be shorter *(44),* in other words, that bonding with fluorine has greater π character. Therefore the carbon-fluorine bond in fluoromethane would be more p-rich and longer than the bond in fluoroform *(45)*. The number of alternative explanations that have been put forward suggests how clouded understanding of these simple experimental observations may be. As might be expected, fluorination also has an effect on carbon-carbon bond strength. The carbon-carbon bond in 1,1,1-trifluoroethane or hexafluoroethane is considerably strengthened relative to that of ethane, 59 or 42 kJ per mole, respectively (*46,47*).

Table I. Carbon-Fluorine Bond Lengths and Bond Dissociation Energies in Halomethanes (*42-43*)

	CH_3X	CH_2X_2	CHX_3	CX_4
F	1.385 Å 456 kJ mol^{-1}	1.358 Å 510 kJ mol^{-1}	1.332 Å 535 kJ mol^{-1}	1.317 Å 543 kJ mol^{-1}
Cl	1.782 Å 350 kJ mol^{-1}	1.772 Å 339 kJ mol^{-1}	1.767 Å 325 kJ mol^{-1}	1.766 Å 301 kJ mol^{-1}
Br	1.939 Å 289 kJ mol^{-1}	1.934 Å 267 kJ mol^{-1}	1.930 Å 259 kJ mol^{-1}	1.942 Å 235 kJ mol^{-1}

Source: Adapted from ref. 22.

Another general experimental phenomenon that is particularly pronounced with organofluorine compounds is the gauche effect, the name given to the tendency of 1,2-difluoroethane to exist in a synclinal conformation as opposed to the antiperiplanar conformation, that would be expected on the basis of dipolar repulsions. Although the gauche effect is a general phenomenon that can occur with any vicinal electronegative substituents, it is most pronounced with vicinally fluorinated molecules. It is has been rationalized by invoking molecular orbital interactions between the fluorines (*48,49*).

The electronegativity of fluorine can also have pronounced effects on the electron distribution within a molecule bearing other substituents and affect the dipole moment of the molecule, the overall reactivity and stability of the compound and acidity or basicity of the neighboring groups. Fluorine can also participate in hydrogen bonds (*50,51*) or function as a ligand for alkali metals because of the available three non-bonded electron pairs (*52*).

Effects of Fluorine on Chemical Reactivity.

Important differences in chemical reactivity of fluorinated compounds are based upon the difference in electronegativity between fluorine and hydrogen, on the higher carbon-fluorine bond strength versus the strength of the carbon-hydrogen bond, and on the ability of fluorine to participate in hydrogen bonding as an electron pair donor. The effects of fluorination have been thoroughly summarized by Chambers (*24*), Smart (*22*) and others (*53*).

Fluorine has a pronounced electron withdrawing effect by relay of an induced dipole along the chain of bonded atoms, a sigma withdrawing effect, I_σ, or this withdrawing may also result from a through space electrostatic interaction also known as a field effect (*54,55*). These effects are apparent when the acidity of trifluoroacetic acid (pKa 0.3) is compared with that of acetic acid (pKa 2.24). However in the gas phase, although fluoroacetic acid is more acidic than acetic acid, it is less acidic than chloroacetic acid, presumably as a result of the lower polarizability of the carbon-fluorine bond, 0.53 $Å^3$, relative to that for the carbon-chlorine bond, 2.61 $Å^3$ (*56*). This diminished charge induced dipole effect of the carbon-fluorine bond, relative to the carbon-chlorine bond, is much closer to the polarizability of the carbon-hydrogen bond. The electronic effect of fluorine directly attached to a π system can be especially complex, as electrons from fluorine may be donated back to the π system in an I_{π} repulsive interaction. These repulsive interactions are most important in the reactions of α–fluorinated anions and radicals and in additions of nucleophiles to fluorinated alkenes.

In contrast Fluorine stabilizes α–cations by the interaction of the vacant p-orbital of the carbocation with the filled orbitals of fluorine. Yet α-trifluoromethyl groups are strongly destabilizing to cationic centers relative to methyl groups. Trifluoromethyl groups have a very pronounced effect in solvolytic studies and have featured prominently in some classic studies (*57*). In recent work, it has been shown that fluorinated carbon-carbon bonds are also capable of directing the stereochemistry of reactions. The more electron deficient carbon-carbon bond (**2a**) has less electron density available to donate to a developing vacant orbital; therefore the stereochemistry of the process is controlled by the more electron rich non-fluorinated carbon-carbon bond (**2b**) (*58-61*).

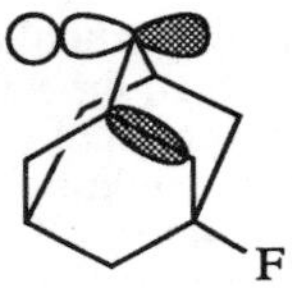

F F

2a 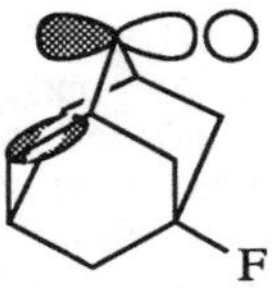 2b

Fluorine bonded directly to a carbanionic center is generally destabilizing by I_{π} repulsive interactions but when the carbanion ic atom is surrounded by trifluoromethyl groups, it is stabilized. In the case of pentakis-(trifluoromethyl)cyclopentadiene (**3**), stabilization of the carbanion accounts for a surprisingly high acidity (pKa<-2), comparable to nitric acid (*62*).

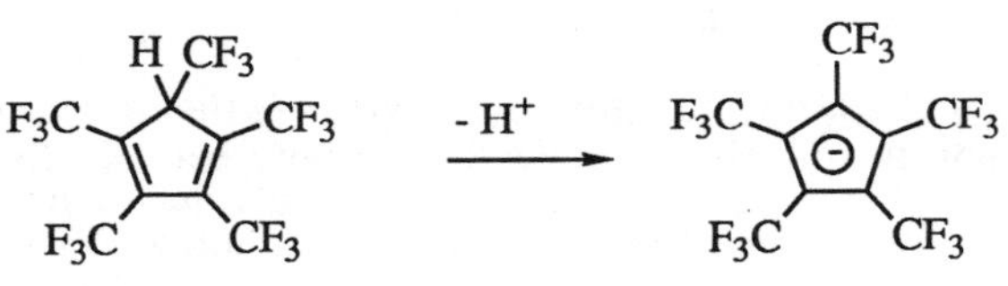

H CF_3 F_3C CF_3 F_3C CF_3 — $-H^+$ → CF_3 F_3C CF_3 F_3C CF_3

3

When speculating about the stability of α-fluorocarbanionic centers, the opposing electron withdrawing effects and I_{π} repulsive effects must be considered. The geometry of the system thus becomes very important in evaluating the magnitude of each of these contributions. Pyramidal α-fluorinated carbanions may be more stable than non-fluorinated pyramidal carbanions; for example, trifluoromethane is significantly more acidic than methane. If, however, fluorine resides on a planar carbanionic center, the ion is less stable; difluoroacetates form enolates less easily than acetates, presumably since the α-proton is less acidic (*63*).

Whereas fluorine bonded to a radical center may have a profound effect on the geometry of that center, the effect of fluorination on the stability of the radical is difficult to assess. However, the increased accessibility of the carbon-fluorine bond LUMO and raised energy of the HOMO have been cleverly employed to direct the selectivity of radical reactions. A developing singly occupied orbital can be stabilized sufficiently by neighboring substituents to direct the stereochemical course of reactions, as in (**4**) (*64*).

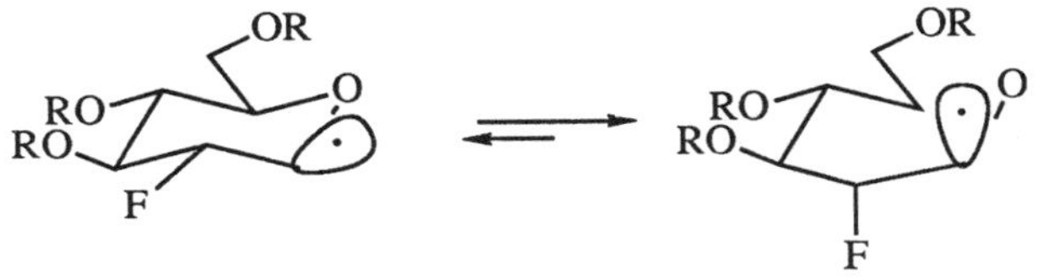

4

Fluorination as a Probe of Biological Reactivity

The attractiveness and utility of fluorine as a substituent in a biologically active molecule results from the pronounced electronic effects which may occur on fluorination combined with the fact that fluorine is not a sterically demanding substituent. With its small Van der Waal's radius (1.35 Å) fluorine closely resembles hydrogen (Van der Waals radius 1.20 Å). On the other hand, the carbon-fluorine bond length, 1.39 Å, is comparable to that of the carbon-oxygen bond length, 1.43 Å, which suggests that such substitutions should have little steric effect on the molecule into which they are made. However, where the Van der Waals volume of a methyl group, using a half sphere approximation, is only 19.6 $Å^3$, that of a trifluoromethyl group is 42.6 $Å^3$ more nearly that of an isopropyl group.

Fluorine can be introduced into a biologically active molecule to block metabolism, to serve as a probe of hydrogen bonding, to function as a leaving group in a enzyme inhibition or as a potent substituent for modifying chemical reactivity of adjacent functionality. Systematic substitution of fluorine for hydroxyl can help establish the effect of hydroxylation on the metabolism of a molecule. This principle has been successfully applied in the study of fluorinated vitamin D_3 analogs. The high carbon fluorine bond strength (108-456 kJ kcal/mole) renders the fluorine substituent resistant to many such metabolic transformations. Fluorine may also be employed as leaving group in addition-elimination processes where its superior leaving group ability relative to hydrogen is important. Such applications have led to the development of very effective mechanism based enzyme inhibitors. As previously mentioned, fluorine can also be a useful probe of hydrogen bonding interactions, since it may act as an electron pair donor but is otherwise unable to participate in such bonding interactions. Lastly, fluorine also has been identified as improving the lipophilicity of a molecule and hence its distribution within an organism.

The natural stable isotope of fluorine, fluorine-19 (^{19}F), with a spin of one-half and a chemical shift range of around 300 ppm, is a sensitive and useful probe in nuclear magnetic resonance (NMR) studies. Fluorine substitution may be a very effective method for studying the fate of bioactive molecules. Since there are few natural fluorinated materials to create background signals, the analyses are freed from the complications often associated with proton NMR spectroscopy (*65*). An artificially prepared useful short-lived isotope, fluorine-18 (^{18}F), decays by positron emission. Positron emission tomography (PET) is an especially useful non-invasive technique for the survey of living tissue which complements traditional methods, such as X-ray studies, by allowing real time analysis of metabolic processes (*66*). Introduction of ^{18}F containing materials into living tissue is an essential part of PET. While other isotopes such as ^{11}C, ^{13}N, or ^{15}O have half lifetimes of twenty, ten and two minutes, ^{18}F has a convenient half life time of 110 minutes, sufficient for synthesis and for administration of the radiolabeled materials (*67*). One application of ^{18}F-positron emission tomography is in the brain imaging of Parkinsonian patients. Using ^{18}F-labeled fluorodopa (**6**), new insights into the chemistry and metabolism of brain have been revealed.

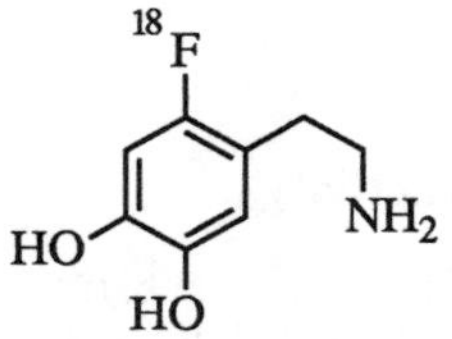

6α-[^{18}F]-Fluoro-L-dopamine

5

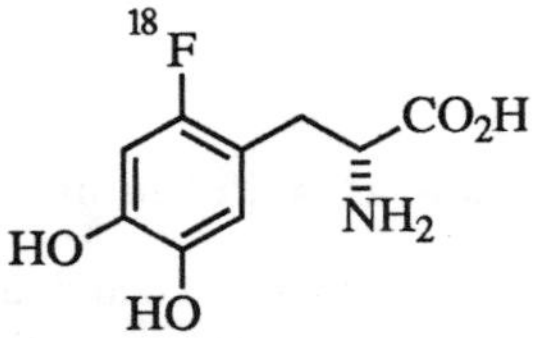

6α-[^{18}F]-Fluoro-L-dopa

6

In yet another example, ^{18}F-labeled estrogen (**7**) may be useful in diagnosing breast tumors by positron emission tomography.

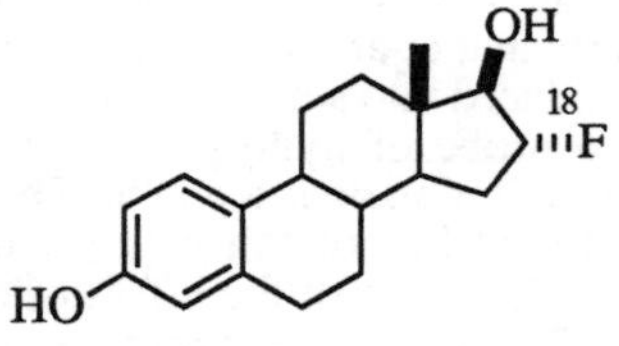

16α-[^{18}F]-17β-Fluoroestradiol

7

Methods for the Introduction of Fluorine

The best source of information on techniques for the introduction of fluorine into organic compounds and for information on preparative methods is the excellent treatise by Hudlicky (*23*), Houben-Weyl's Methoden der Organischen Chemie(*68*), as well as the books of Shephard and Sharts (*25*) and Chambers (*24*). More recently, methods for the fluorination of organic molecules have been reviewed by several authors (*15,69-75*). More specialized reviews discuss the synthesis of α–fluorinated carbonyl compounds (*76*), which appear quite often in biologically active molecules. An especially rapidly growing area of interest is fluorination with molecular fluorine or with reactive species prepared by means of molecular fluorine (*77-81*). Methods for the construction of more exhaustively fluorinated molecules have also been recently collated (*26*). The utility of selectively fluorinated molecules as enzyme substrates (*82-83*) has been reviewed as has the preparation of fluorinated analogs of insect juvenile hormones and pheromones (*84*). Also progress reports on the preparation and biological activity of fluorinated analogs of vitamin D3 (*85*), on the preparation and biochemistry of fluorinated carbohydrates (*86*) and of fluorinated amino acids have appeared in the literature (*87-88*).

Fluorination of Aminoacids and Amines. Fluorinated analogs of naturally occurring aminoacids and amines exhibit unique physiological activities. A number of fluorine containing amino acids have been synthesized and studied as potential enzyme inhibitors and therapeutic agents (*89*). β-Fluorine substituted amino acids are generally regarded as potential irreversible inhibitors of pyridoxal phosphate-dependent enzymes(*89a,b-95*). Fluorinated analogs are accepted by enzymes as substrates, as fluorine is comparable in steric demand to hydrogen, but often are not substrates for normal enzymatic transformation.

α-Monofluoromethyl and difluoromethyl amino acids have been recognized as potent enzyme-activated irreversible inhibitors of parent α-amino acid decarboxylases. Examples of the physiologically important fluorinated amines formed by decarboxylation are dopamine (**8**), 5-hydroxytryptamine (serotonin) (**9**), histamine (**10**), tyramine (**11**) and gamma-aminobutyric acid (GABA) (**12**) (*82*). The catechol amines are important in peripheral and central control of blood pressure (*96*). Elevated histamine levels are observed in diseases such as allergies, hypersensitivity, gastric ulcers and inflammation (*97*). High putrescine levels are associated with rapid cell development, including tumor growth (*98*).

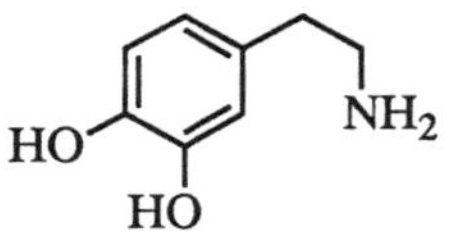

8 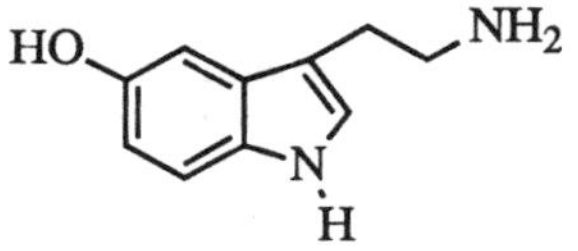 9 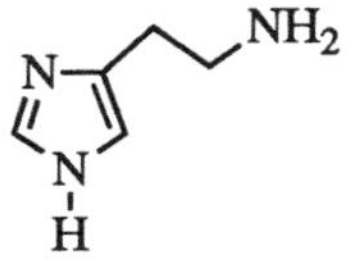 10

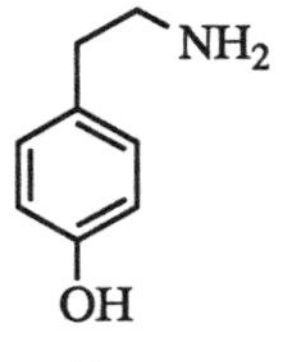

11 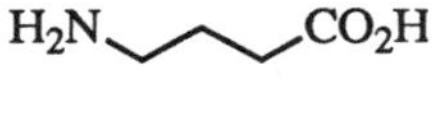 12

Aminoacids containing the trifluoromethyl group are also potential antimetabolites (*99*). Methyl groups may be replaced by trifluoromethyl groups in the preparation of amino acid analogs. The high electron density of the trifluoromethyl group may be important in the formation of strong hydrogen bonds with enzymes, thereby blocking enzymatic metabolism of the natural substrates. This group is also attractive since it is relatively non-toxic and somewhat more stable than the mono- and difluoromethyl analogs.

The synthesis of fluorinated analogs of biologically important amines has been explored extensively. Their direct preparation is difficult since most of the common fluorination agents can also react with the amino group. However, by the use of appropriate solvents, like liquid hydrogen fluoride which protects the amine functionality by protonation, successful fluorinations have been achieved (*100*). β–Fluorinated amines are important targets in the design of antimetabolites and drugs again since fluorine causes minimal structural changes and maximal shift in electron distribution.

Fluorinated Carbohydrates. Selectively fluorinated carbohydrates have found utility in probing biochemical mechanisms or in modifying the activity of glycosides (*86, 101-104*). These materials have many applications in biochemistry, medicinal chemistry and pharmacology. Only one naturally occurring fluorinated carbohydrate is known (*86*), the 4-deoxy-4-fluorosugar constituent of nucleocidin (**13**).

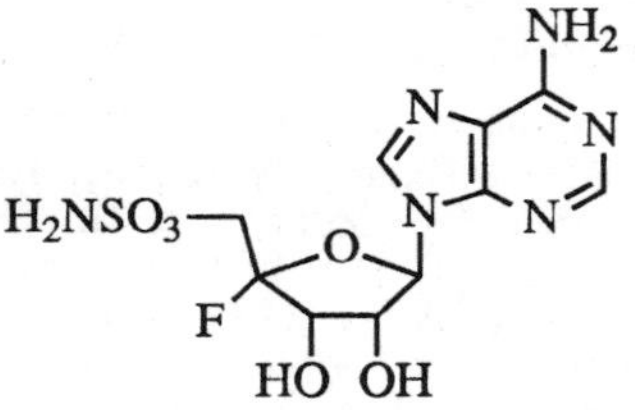

NH_2 N N N N H_2NSO_3 O F HO OH

13

The biochemical rationale for incorporating fluorine in the carbohydrate residue is that replacement of a hydroxyl by fluorine would cause only a very minor steric perturbation of the structure or conformation while at the same time would have a profound electronic effect on neighboring groups. The substitution is possible while retaining the capacity of the position as an acceptor in hydrogen bonding. Yet these same attributes make the synthesis of fluorinated carbohydrates difficult. The synthesis of fluorinated carbohydrates offers a particularly fruitful field for the combination of modern chemical and enzymatic synthetic techniques. Total synthesis would be difficult because of the stereochemical control required at the multiple adjacent asymmetric centers of a fluorinated carbohydrate (*105*).

Fluorinated Analogs of Nucleic Acids. Fluorinated analogs of the naturally occurring nucleic acids have become established as antiviral (*106*), antitumor (*107,108*) and antifungal agents. Many fluorinated nucleosides exhibit biological activity in as much as the structures differ only slightly sterically from the naturally occurring molecules. The electronic effects of the fluorine substituent play a major role in the biological activity of the analogs. Fluorine has been employed as a replacement for both hydroxyl and hydrogen and difluoromethylene units have been employed as replacements for oxygen. The difluoromethylene group is only slightly larger than an oxygen atom, in fact there is not another functional group which can replace matches so well the steric and electronic demand of oxygen.

Fluorinated Aromatics. Fluorinated aromatics have found wide use as antibiotics, sedatives, important agrochemicals and radiochemical imaging agents. The best method is of course diazotization of an aromatic amine followed by dediazoniation in the presence of fluoride-containing counterion. Selective electrophilic fluorination procedures are rare *(109)*. Unlike other halogens, fluorine cannot be directly introduced into specific positions of aromatic compounds using elemental fluorine. Most syntheses of bioactive materials start therefore with unfunctionalized fluorinated aromatic materials upon which the functionality is later arrayed.

Fluorine Substitution of Prostanoids. Fluorination of prostaglandins, prostacyclins and thromboxanes has led to exciting and useful modifications of activity. The biosynthetic pathway of these compounds begins with arachidonic acid (**14**).

arachidonic acid
14

PGH$_2$

PGF$_2$

TXA$_2$

PGI$_2$

Selective fluorination of prostanoids has been effected on the cyclopentane nucleus and on both side chains. A variety of techniques have been employed but may be arbitrarily divided into the fluorination reactions of intermediates and the use of fluorinated building blocks to prepare the target compounds. The chemistry and biology of fluorinated prostaglandins, prostacyclins and thromboxanes through 1981 has been reviewed (*110*).

Fluorinated Steroids. The fluorination of steroids has been known to have profound effects on biological activity since the early work of Fried (*40*). Fluorinated steroids have been described in other reviews (*111*). The use of fluorination as a tool to enhance biological selectivity and to improve the utility of a biologically active material has been clearly demonstrated in studies of vitamin D$_3$.

The preparation of fluorinated analogs of vitamin D$_3$ has been reviewed (*112-114*). Fluorination has been employed to prevent metabolic hydroxylation or to modify the reactivity of hydroxyl groups. Vitamin D_3 (**15**) is activated by hydroxylation at C-1 and C-25 to yield the active steroid hormone (**16**).

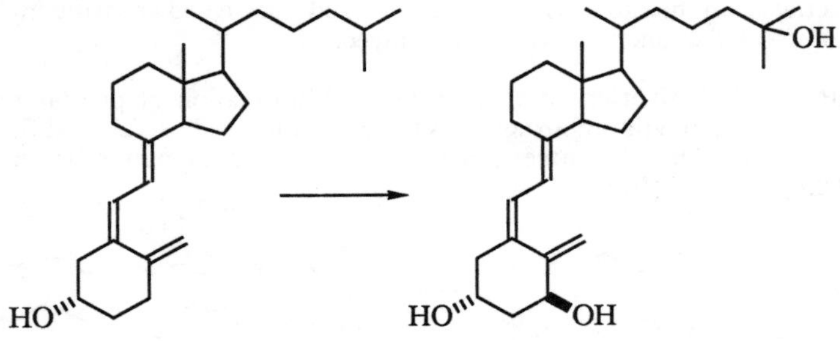

15 16

Fluorination at C-25 (**17**), C-1 (**18**), or C-3 (**19**) prohibited the hydroxylation reaction that was essential for activation, and facilitated a study of the role each of these hydroxyl groups played in the various aspects of the activity of vitamin D_3.

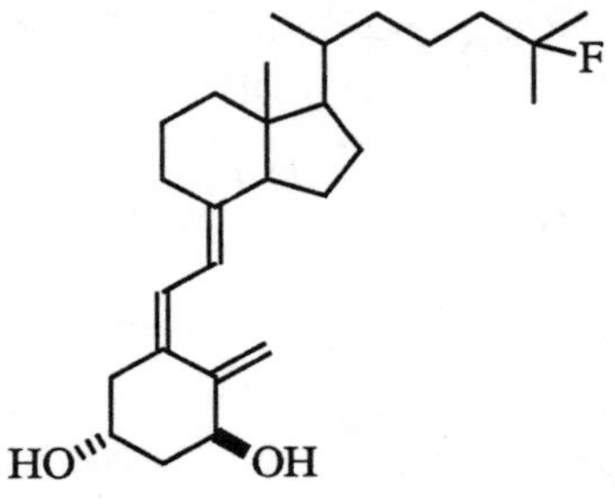

17 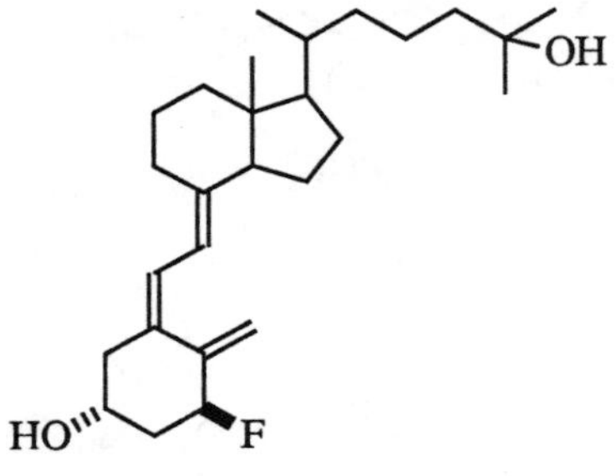 18

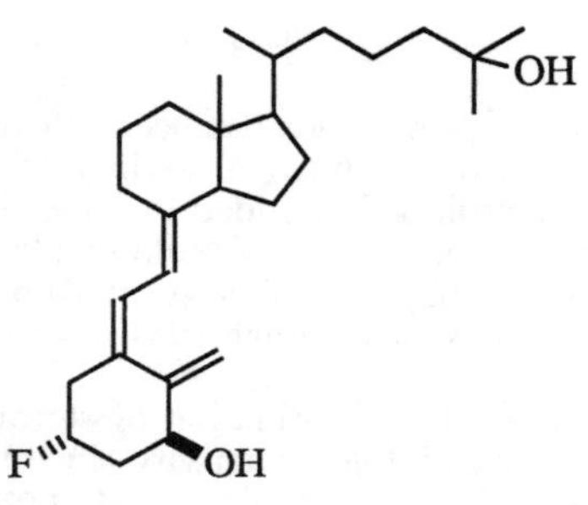

19

Compounds which were fluorinated at C-24 (**20**), C-26 (**21**), or C-23 (**22**), prevented the hydroxylation reactions that lead to metabolic deactivation of the vitamin D_3.

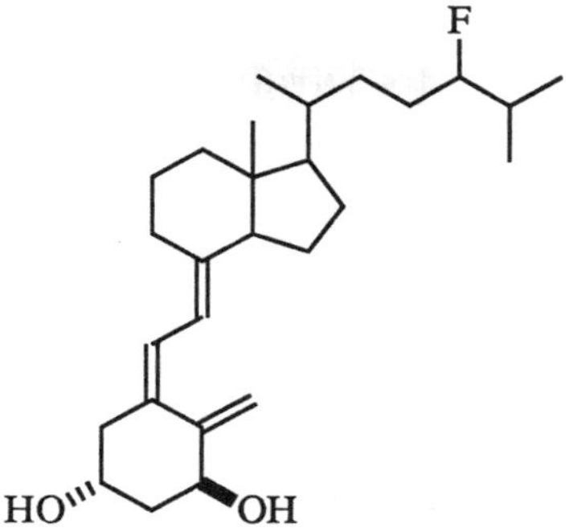

F
HO''' OH
20

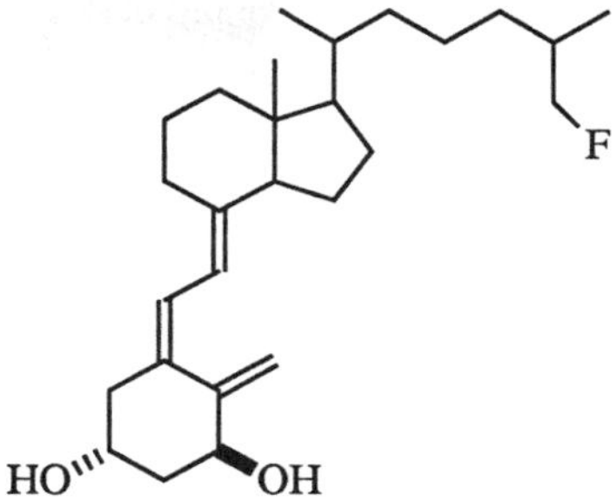

F
HO''' OH
21

Further, fluorination at C-24 (**23**), C-26 (**24**), or C-2 (**25**), where the fluorine would be adjacent to the 25- or 1- hydroxyls essential for activity, was useful in determining the role of those hydroxyls in hydrogen bonding, either as hydrogen bonding donors or acceptors.

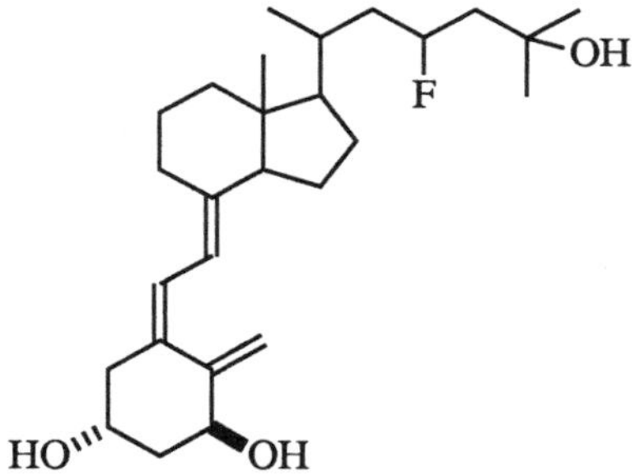

OH
F
HO''' OH
22

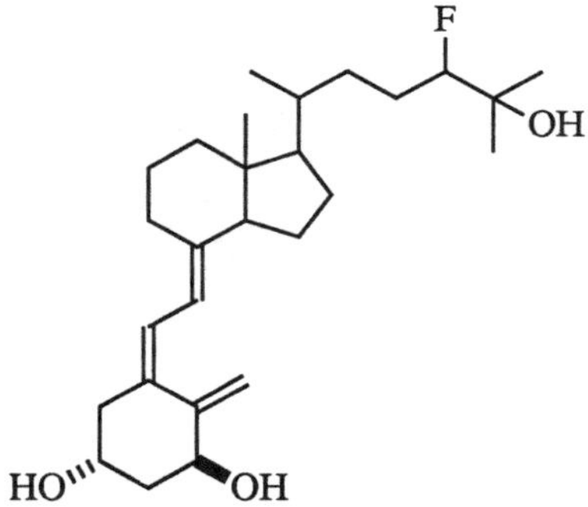

F
OH
HO''' OH
23

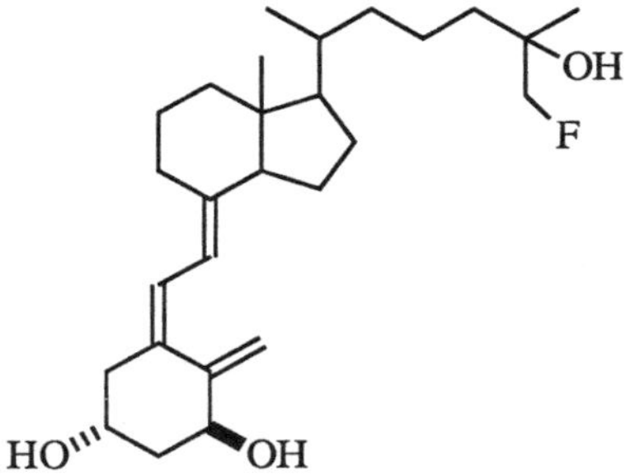

OH
F
HO''' OH
24

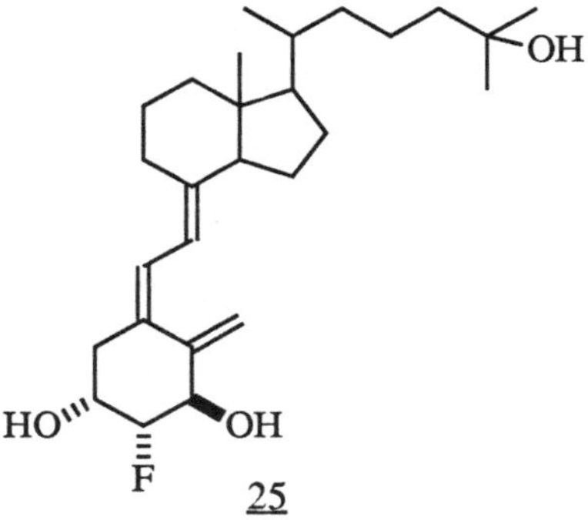

OH
HO''' OH
F
25

Additionally, fluorination was used to study the photochemical reaction which forms vitamin D_3. Both the 6-fluoro and 19,19-difluoro analogs,

(**26**) and (**27**), respectively, where fluorination substitution of the reactive olefin system has been made, were prepared.

F

HO

26

F_2HC

OH

27

Although the scope of the utility of fluorination in modifying activity is clearly illustrated in the above examples, this utility is not limited to vitamin D derivatives alone.

Conclusion

This small survey of the uses of fluorination in selected substrates illustrates clearly the potential selective fluorination has as a tool to modify the reactivity and biological activity of those substances. Better and more effective understanding of chemistry of selectively fluorinated materials must necessarily await further evolution of not only synthetic methods but also of the underpinning chemical theory. Some latest advances in theory, synthetic methods and applications are chronicled in this volume on the subsequent pages.

References

1. Ács, M.; V. D. Busche, C.; Seebach, D. *Chimia* **1990**, *44*, 90.
2. Goldman, P. *Science* **1969**, *164*, 1123.
3. Smith, F. A. *Chem. Tech.* **1973**, 422.
4. Filler, R. *Chem. Tech.* **1974**, 752.
5. *Ciba Foundation Symposium, Carbon-Fluorine Compounds, Chemistry, Biochemistry and Biological Activities*. Elsevier: New York, NY **1972**.
6. Filler, R. In *Organofluorine Chemicals and Their Industrial Applications;* Banks, R. E., Ed.; E. Horwood: Chichester, **1979**.
7. *Biomedical Aspects of Fluorine Chemistry*, Filler, R.; Kobayashi, Y., Ed.; Kodansha Ltd.: Tokyo, **1982**.
8. *Biochemistry Involving Carbon-Fluorine Bonds;* Filler, R., Ed.; American Chemical Society: Washington, D.C., **1976**.
9. Gerstenberger, M. R. C.; Haas, A. *Angew. Chem. Intl. Ed. Engl.* **1981**, *20*, 647.
10. Welch, J. T. *Tetrahedron* **1987**, *43*, 3123.
11. *Synthesis and Reactivity of Fluorocompounds*; Ishikawa, N., Ed.; CMC: Tokyo, **1987**, Vol. 3.
12. *Preparation, Properties and Industrial Applications of Organofluorine Compounds* ; Banks, R. E., Ed.; E. Horwood: Chichester, **1982**.
13. Kumadaki, I. *J. Synth. Org. Chem. Jap.* **1984**, *42*, 786.

14. Bannai, K.; Kurozumi, S. *J. Synth. Org. Chem. Jap.* **1984**, *42*, 794.
15. Mann, J. *J. Chem. Soc. Rev.* **1987**, *16*, 381.
16. Fujita, M.; Hiyama, T. *J. Synth. Org. Chem. Jap.* **1987**, *45*, 664.
17. Kitazume, T.; Yamazaki, T. *J. Synth. Org. Chem. Jap.* **1987**, *45*, 888.
18. Ojima, I. *L'Actualité Chimique*, **1987**, *May*, 179.
19. Imperiali, B. *Biotechnol. Processes, 10 Synth. Pept. Biotechnol.* **1988**, 97.
20. Bey, P. *Actual. Chim. Ther.-16eSerie* **1989**, 111.
21. Shimizu, M.; Yoshioka, H. *J. Synth. Org. Chem. Jap.* **1989**, *47*, 27.
22. Smart, B. In *Chemistry of Functional Groups, Supplement D, The Chemistry of Halides, Pseudohalides and Azides*; Patai, S.; Rapoport, Z. Eds.; J. Wiley: New York, NY, **1983**, Suppl. D; pp. 603-655.
23. Hudlicky, M. *Chemistry of Organic Fluorine Compounds, 2nd Ed.;* E. Horwood: Chichester, **1976**.
24. Chambers, R. D. *Fluorine in Organic Chemistry;* Wiley-Interscience: New York, NY, **1973**.
25. Sheppard, W. A.; Sharts, C. M. *Organic Fluorine Chemistry;* W. A. Benjamin: New York, NY, **1968**.
26. *Synthesis of Fluoroorganic Compounds;* Knunyants, I. L.; Yacobson, G. G., Eds.; Springer-Verlag: New York, **1985**.
27. *Fluorine Containing Molecules;* Liebman, J. F.; Greenberg, A.; Dolbier, Jr., W. R., Eds.; VCH: Deerfield Beach, FL, **1988**.
28. Welch, J. T.; Eswarakrishnan, S. *Fluorine in Bioorganic Chemistry;* J. Wiley: New York, NY, 1990.
29. Moissan, H. *Compt. Rend.* **1886**, *102*, 1543; *103*, 203.
30. Swarts, F. *Bull. Acad. Roy. Belg.* **1892**, *24*, *[3]*, 474.
31. Midgley, T.; Henne, A. L. *Ind. Eng. Chem.* **1930**, *22*, 542.
32. McBee, E. T.; et. al. *Ind. Eng. Chem.* **1947**, *39*, 235.
33. *Fluorine Chemistry;* Simons, J. H., Ed.; Academic Press: New York, NY, Vol 1, 1950.
34. *Preparation, Properties and Technology of Fluorine and Organic Fluorine Compounds;* Slesser, C.; Schram, S. R., Eds.; McGraw-Hill: New York, NY, 1951.
35. Simons, J. H. *J. Electrochem. Soc.* **1949**, *95*, 47.
36. Simons, J. H.; Francis, H. T.; Hogg, J. A. *J. Electrochem. Soc.* **1949**, *95*, 53.
37. Simons, J. H.; Harland, W. J. *J. Electrochem. Soc.* **1949**, *95*, 55.
38. Kauck, E. A.; Diesslin, A. R. *Ind. Eng. Chem.* **1951**, *43* 2332.
39. Simons, J. H.; Pearlson, J. H.; Brice, W. H.; Watson, W. A.; Dresdner, R. D. *J. Electrochem. Soc.* **1949**, *95*, 59.
40. Fried, J.; Sabo, E. F. *J. Am. Chem. Soc.* **1954**, *76*, 1455.
41. Yokozeki, A.; Bauer, B. *Topics in Curr. Chem.* **1975**, *53* , 71.
42. Patrick, C. R. *Adv. Fluorine Chem.* **1961**, *2*, 1.
43. Egger, K. W.; Cooks, A. T. *Helv. Chim. Acta* **1973**, *56*, 1516.
44. Peters, D. *J. Chem. Phys.*, **1963**, *38*, 561.
45. Bent, H. A. *Chem. Rev.* **1961**, *61*, 275.
46. Rodgers, A. S.; Ford, A. G. *Int. J. Chem. Kinetics* **1973**, *99*, 691.
47. Chen, S. S.; Rodgers, A. S.; Chow, J.; Wilhoit, R. C.; Zwolinski, B. J. *J. Phys. Chem. Ref. Data* **1975**, *4*, 441.
48. Friesen, D.; Hedberg, K. *J. Am. Chem. Soc.* **1980**, *102*, 3987.
49. Fernholt, L.; Kveseth, K. *Acta Chem. Scand. A.* **1980**, *43*, 163.
50. Baker, A. W.; Shulgin, A.T. *Nature(London)* **1965**, *206*, 712.

51. Doddrell, S.; Wenkert, E; Demarco, P. V. *J. Mol. Spec.* **1969**, *32*, 162.
52. Carrell, H. L.; Glusker, J. P.; Piercy, E. A.; Stallings, W. C.; Zacharias, D. E.; Davis, R. L.; Astbury, C.; Kennard, C. H. L. *J. Am. Chem. Soc.* **1987**, *109*, 8067.
53. *L'Actualité-Chimique* **1987**, *May*, 135-188.
54. Topson, R. D. *Prog. Phys. Org. Chem.* **1976**, *12*, 1.
55. Levitt, L. S.; Widing, H. F. *Prog. Phys. Org. Chem.* **1976**, *12*, 119.
56. Fraga, A.; Saxena, K. M.; Lo, B. W. N. *Atomic Data* **1971**, *3*, 323.
57. Brown, H. C.; Kelly, D. P.; Periasamy, M. *Proc. Natl. Acad. Sci.* **1980**, *77*, 6956.
58. Lin, M.-H.; Boyd, M. K.; LeNoble, W. J. *J. Am. Chem. Soc.* **1989**, *111*, 8746.
59. Lin, M.-H.; Cheung, C.-K.; LeNoble, W. J. *J. Am. Chem. Soc.* **1988**, *110*, 6652.
60. Cheung, C.-K.; Tseng, L. T.; Lin, M.-H.; Srivastava, S.; LeNoble, W. J. *J. Am. Chem. Soc.* **1986**, *108*, 1598.
61. Xie, M.; LeNoble, W. J. *J. Org. Chem.* **1989**, *54*, 3839.
62. Laganis, E. D.; Lemal, D. M. *J. Am. Chem. Soc.* **1980**, *102*, 6633.
63. Modena, G.; Scorrano, G. In *The Chemistry of the Carbon Halogen Bond;* Patai, S., Ed.; Wiley: London, 1973, pp 301-406.
64. Giese, B. *Pure and Applied Chem.* **1988**, *60*, 1655-1658.
65. Gehrig, J. T. In *Methods in Enzymology;* Openheimer, N. J.; Jais, T. L. Eds.; Academic Press: San Diego, 1989, pp 3-22.
66. *Chemical and Engineering News,* Aug. 15, 1988, pp 26-29.
67. Reivich, M.; Alavi, A. *Positron Emission Tomography;* Alan R. Liss: New York, 1985.
68. Forche, E.; *Methoden der Organischen Chemie (Houben-Weyl);* Müller, E., Ed.; Georg Thieme: Stuttgart, 1962, Vol. 5/3.
69. Rozen, S. *Acc. Chem. Res.* **1988**, *21*, 307.
70. Boswell, G. A.Jr., Ripka, W. C., Scribner, R. M., Tullock, C. W. *Org. Reactions* **1974**, *21*, 1.
71. Sharts, C. M. *Org. Reactions* **1974**, 21, 125.
72. Wang, C.-L. J. *Org. Reactions* **1985**, *34*, 319-400.
73 Hudlicky, M. *Org. Reactions* **1987**, *35*, 513-637.
74. *New Fluorinating Agents In Organic Synthesis;* German, L.; Zemskov, S., Eds.; Springer: Berlin, 1989.
75. Bohlmann, R. *Nachr. Chem. Tech. Lab.* **1990**, *38*, 40-43.
76. Rozen, S.; Filler, R. *Tetrahedron* **1985**, *41*, 1111.
77. Gerstenberger, M. R. C.; Haas, A. *Angew. Chem.* **1981**, *93*, 659.
78. Vyplel, H. *Chimia* **1985**, *39*, 305.
79. Haas, A.; Lieb, M. *Chimia* **1985**, *39*, 134.
80. Purrington, S.; Kagen, B. S.; Patrick, T. B. *Chem. Rev.* **1986**, *86*, 997.
81. Auer, K.; Hungerbühler, E.; Lang, R. W. *Chimia* **1990**, *44*, 120-123
82. Walsh, C. *Tetrahedron* **1982**, *38*, 871-909.
83. Walsh, C. In *Adv. in Enzymology;* Meister, A., Ed.; John Wiley and Sons: New York, 1983, Vol. 55, p 197.
84. Prestwich, G. D. *Pesticide Science* **1986**, *37*, 430.
85. Ishikawa, N. *Kagaku To Seibatsu* **1984**, *22*, 93.
86. *Fluorinated Carbohydrates, Chemical and Biochemical Aspects;* Taylor, N. F., Ed.; American Chemical Society: Washington, D.C., 1988.

87. Loncrini, D. F., Filler, R. *Adv. Fluorine Chem.* **1970**, 6, 43.
88. Uchida, K.; Tanaka, H. *J. Syn. Org. Chem Jap.* **1988**, *46*, 977-985.
89. Kollonitsch, J.; Perkins, L. M.; Patchett, A. A.; Doldouras, G. A.; Marburg, S.; Duggan, D. E.; Maycock, A. L.; Aster, S. D. *Nature* **1978**, *274*, 906.
90. Abeles, R. H.; Maycock, A. L. *Acc. Chem. Res.* **1976**, *9*, 313-319.
91. Rando, R. R. *Enzyme Inhibitors* **1975**, *8*, 281-288.
92. Metcalf, B. W. *Ann. Rep. Med. Chem.* **1981**, *16*, 289-297.
93. Bey, P. *Ann. Chim. Fr.* **1984**, *9*, 695-702.
94. Unbeless, J. C. Goldman, P. *Mol. Pharmacol.* **1970**, 6, 46; 1971, 7, 293.
95. Licato, N. J., Coward, J. K. Nimec, Z., Galivan, J. Bolanowska, W. E., McGuire, J. J. *J. Med. Chem.* **1990**, 33, 1022.
96. Versteeg, D. H. G.; Palkouts, M.; Van der Gugten, J.; Wijnen, H. J. L. M.; Smeets, G. W. M.; DeJong, W. *Prog. Brain Res.* **1977**, *47*, 111.
97. Douglass, W. W. In *The Pharmacological Basis of Therapeutics*; 5th Ed.; Goodman, L. S.; Gilman, A., Eds.; MacMillan: New York, NY, 1975; pp. 590-629.
98. Russell, D. H. In *Polyamines in Normal and Neoplastic Growth* ; Russell, D. H., Ed.; Raven: New York, NY, 1973, pp. 1-13.
99. Walborsky, H. M.; Baum , M.; Loncrini, D. F. *J. Am. Chem. Soc.* **1955**, *77*, 3637-3640.
100. Kollonitsch, J.; Marburg, S.; Perkins, L. M. *J. Org. Chem.* **1979**, *44*, 771-777 .
101. *J. Carbohydr. Res.* **1985**, *4* (special Fluorocarbohydrate issue).
102. Penglis, A. A. E. *Adv. Carbohydr. Chem. Biochem.* **1981**, *38*, 195.
103. Foster, A. B.; Westwood, J. H. *Pure Applied Chem.* **1973**, *35* , 147.
104. Kent, P. W. *Chem. Ind.* **1969**, 1128.
105. Zamojski, A.; Banaszek, A.; Grynkiewcz, G. *Adv. Carbohydr. Chem. Biochem.* **1982**, *40*, 1-20.
106. Bergstrom, D. E.; Swartling, D. J. In *Fluorine Containing Molecules. Structure, Synthesis and Applications;* Liebman, J. F.; Greenberg, A.; Dolbier, W. R., Jr., Eds.; VCH: Deerfield Beach, 1988.
107. Lucey, N. M.; McElhinney, R. S. *J. Chem. Res. (S),* **1985**, 240.
108. Hoshi, A. In *Fluoropyrimidines in Cancer Therapy*; Kimura, K.; Fujii, S.; Ogawa, M.; Bodey, G. P.; Alberto, P., Eds.; Elsevier Science: 1984.
109. Singh, S.; Desmarteau, D. D.; Zyber, S. S.; Witz, M.; Huang, H.-N. *J. Am. Chem. Soc.* **1987**, *109*, 7194-7196
110. Barnette, W.E. *Critical Rev. Biochem.* **1984**, *15*, 201.
111. Wettstein, A. In *Ciba Foundation Symposium, Carbon-Fluorine Compounds, Chemistry, Biochemistry and Biological Activities;* Elsevier: New York, NY, 1972.
112. Kobayashi, Y.; Taguchi, T. In *Biomedical Aspects of Fluorine Chemistry;* Filler, R.; Kobayashi, Y., Eds.; Kodansha Ltd.: Tokyo, 1982.
113. Kobayashi, Y.; Taguchi, T. *J. Syn. Org. Chem. Jap.* **1985**, *43*, 1073.
114 . Ikekawa, N. *Medicinal Chemistry Reviews* **1987**, *7*, 333-366.

RECEIVED October 1, 1990

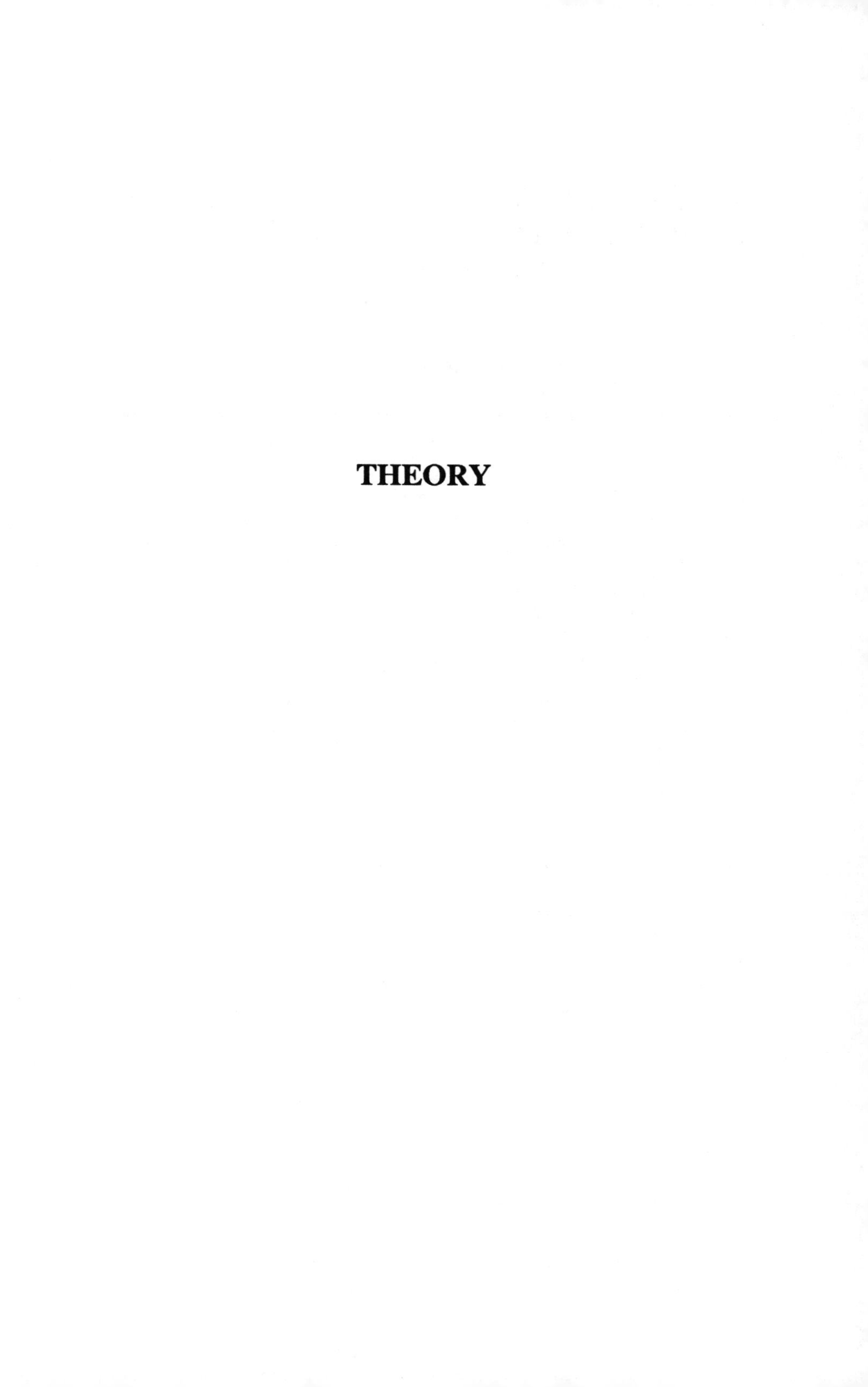

THEORY

Chapter 2

The Effect of Fluorination on Polyacetylene and the Role of Internal Hydrogen Bonds to Fluorine

Molecular Orbital Models

David A. Dixon and Bruce E. Smart

Central Research and Development Department, Experimental Station, E.I. du Pont de Nemours and Company, Wilmington, DE 19880–0328

Molecular orbital calculations on fluorinated butadienes and hexatrienes were used to model the effects of fluorination on the properties of poly(acetylene). Like poly(acetylene), "head-to-head" poly(fluoro- acetylene), (-CH=CF-CF=CH-), is predicted to adopt a planar, all *trans* structure, but poly(difluoro-acetylene) favors a non-planar skewed chain conformation. "Head-to-tail" poly(fluoroacetylene), (-CH=CF-CH=CF-) is predicted to favor a nearly planar *cis* structure stabilized by intramolecular CF···HC hydrogen binding. Calculations on 2-fluoroethanol and on both 2-fluoroacetaldehyde enol and its alkali metal (Li, Na, K) enolates reveal moderately strong intramolecular CF···HO hydrogen bonds(1.9 and 3.2 kcal/mol, respectively) and even stronger intramolecular coordination of CF to alkali metal cations (9-12 kcal/mol).

The presence of a fluorine substituent can dramatically affect molecular properties (*1,2*). The C–F bond length is longer on average than a C–H bond by about 0.25 Å, and the van der Waals radius of F is 1.47 Å as compared with the values of 1.75 Å for Cl and 1.20 Å for H (*3,4*). Thus, F is rather small in size, but it is still significantly larger than H. (Actually, F is very nearly isosteric to O, whose van der Waals radius is 1.52 Å.) Fluorine also is much more electronegative than C; consequently, it always will be negatively charged when bonded to carbon. This results in strong C–F bond dipoles that can significantly affect molecular structure. Furthermore, the carbons bonded to fluorine become positive when more than one fluorine is attached, which can affect the bonding between the carbon atoms. The lone pairs on fluorine can significantly mix with any π orbitals on an adjacent carbon, which also can affect electronic properties, although in simple, planar ethylenes the energy of the HOMO is essentially independent of the number of fluorine substituents (*5*).

Note: This chapter is contribution Number 5651.

0097–6156/91/0456–0018$06.00/0

Although there are many elegant experimental approaches to the study of fluorine substituent effects as outlined elsewhere in this book, our approach is to employ modern theoretical methods, notably *ab initio* molecular orbital theory (*6*), to study these effects. The use of molecular orbital theory to study fluorine was initially hampered by the well-known result that F_2 is unbound with respect to two F atoms at the Hartree-Fock level (*7*) although one can minimize the geometry and find a diatomic structure. Small basis set calculations had been used to predict the structures of simple fluorocarbons, but in contrast to the usual results at the Hartree-Fock level, the C–F bond distances were always too long compared to the experimental values. These results suggested that fluorine would be very difficult to treat as a substituent. Because of Du Pont's long-standing interest in fluorinated materials, we set out to find a computational level that could reliably predict the structures and energetics of fluorinated compounds. We discovered that good geometries could be predicted for fluorinated materials if a double-zeta basis set augmented with polarization functions on carbon was employed for systems containing only C, F, and H. Addition of polarization functions to F and H did little to improve geometries, but in some cases it did improve the accuracies of the calculated reaction energies (*8-11*).

The ability to do these calculations has relied on developments in several areas. The availability of supercomputer cycles has been very important to our studies, but other developments have played key roles as well (*11-14*) [see Figure 1]. Improvements in vectorizing compilers and operating systems have provided a stable computing environment of the kind required to support the large scale simulations that we are undertaking. Moreover, there have been advances in theoretical methods, especially the ability to do analytic first and second derivative calculations, and to calculate correlation corrections on large systems using the MP-2 method. The derivatives are required to perform geometry optimizations in a reasonable way, especially for asymmetric molecules, and to calculate the molecular force fields necessary to confirm that a structure is an energy minimum. Finally, there have been significant improvements in visualization techniques which help to better understand the results of the calculations.

In this chapter, we focus on the effect of fluorine as a substituent in a simple polymeric system, polyacetylene. Polyacetylene, of course, has several potentially practical uses because of its conducting and opto-electronic properties (*15*) and we are interested in studying how F substitution might influence these properties. Our model systems are butadiene and hexatriene, and we discuss both partially fluorinated and perfluorinated materials. Because we discovered that CF···HC hydrogen bonding is important in these systems, we also present results on the nature of the intramolecular hydrogen bond between the CF and OH groups in alcohols and enols. Related results on intramolecular coordination of alkali metals to C–F bonds in fluoroenolates are briefly described.

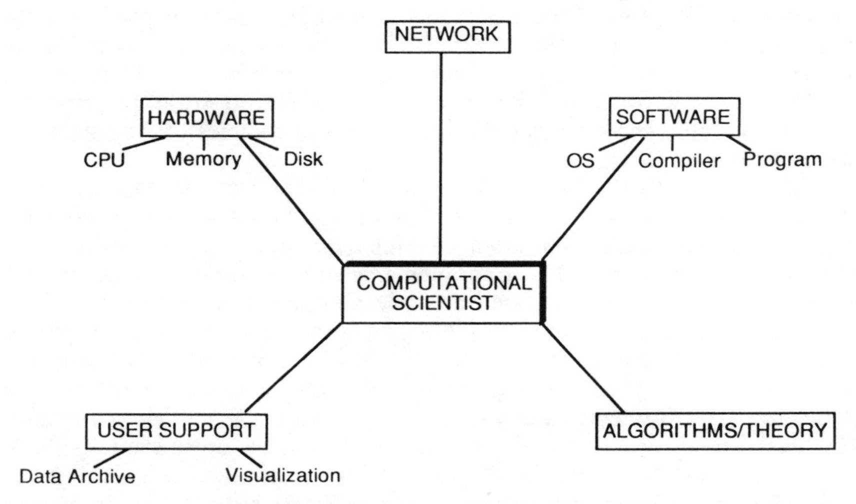

Figure 1. Block diagram of requirements for a computational scientist. (OS = operating system).

Calculations

The calculations discussed below were all done with the program system GRADSCF (*16*) on Cray computer systems. Geometries were optimized using gradient techniques at the SCF level (*17-19*) and force fields (*20, 21*) were calculated for the final structures as were the MP-2 corrections to the energy (*22, 23*). The calculations on the models for substituted acetylenes were all done with a double-zeta basis set, which is flexible in the valence space, augmented by a set of 6 d cartesian polarization functions on carbon (*24*). For the largest calculations discussed below (C_6F_8), there were 162 contracted functions. For the calculations on models for internal hydrogen-bonding involving OH bonds, a triple-zeta basis set (*25*) augmented by two sets of d polarization functions on C, O and F, and a single set of p polarization functions on H was used (*26*). Extended basis sets taken from our previous work on alkali systems were used for the alkali binding studies (*27, 28*).

1,3-Butadiene. 1,3-Butadiene has a well-established, planar *trans* (τ = 180°) structure (*29*). A second *cis*-skew structure has been predicted to exist with a torsion angle (τ) of 37-40° (*30, 31*). The *cis*-skew structure is calculated to be 3.15 kcal/mol above the *trans*, and the experimental estimate of this difference is 2.5 kcal/mol (*32, 33*). The two *cis*-skew structures are separated by a barrier through the *cis* conformer of 0.5 kcal/mol (*22, 23*). Perfluoro-1,3-butadiene, however, has the *cis*-skew structure as the global minimum with $\tau = 47.4 \pm 2.4°$ from electron diffraction measurements (*34*). From photoelectron and UV spectral data, τ is estimated to be $42 \pm 15°$ (*35*). Our *ab initio* calculations gave a comparable value of 58.4° (*10*). Clearly, fluorine substitution substantially affects the gross structure of the butadiene.

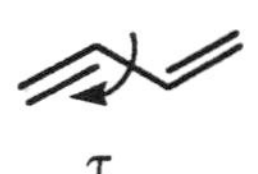

τ

Table I. Torsional Potential Energy Surface for Perfluoro-1,3-butadiene

τ^a	$\Delta E(SCF)^b$	$\Delta E(MP\text{-}2)^b$	ν^c
0	5.58	6.26	77i
60	0.00	0.00	43
123	1.07	0.71	12i
133	1.05	0.79	25
180	1.77	2.48	30i

[a]Torsion angle in degrees; *cis* = 0°, *trans* = 180°. [b]Relative energies in kcal/mol. [c]Torsional frequencies in cm^{-1}.

At the time the original theoretical studies were done, we were able to optimize only three points for the torsion potential and to do frequency calculations on the minimum energy structure only at the 6-31G*(C) level. Since then we have completed a more detailed study of the potential energy surface for torsion about the C_2–C_3 bond. A molecular graphics view of the minimum energy structure is shown in Figure 2 (*36*). The twist about the C_2–C_3 bond is 60°. The various energies are shown in Table I and a plot of the torsional potential is given in Figure 3. The SCF results show that there are two minima on the torsion surface, the global *cis*-skew minimum and a local *trans*-skew minimum with $\tau = 133°$. The *trans*-skew minimum is above the *cis*-skew minimum by 1.05 kcal/mol, and the surface is very flat in this region as evidenced by the low torsion frequency of only 25 cm^{-1}. The barrier in going from the *trans*-skew to the *cis*-skew is only 0.02 kcal/mol (7 cm^{-1}) suggesting that the *trans*-skew minimum probably does not support a zero-point level consistent with the very flat nature of the surface. The transition state between the two skew structures is at $\tau =$ 123°, and the imaginary frequency is very low, only 12i cm^{-1}. The *cis* and *trans* structures are at higher energies, 5.6 and 1.8 kcal/mol, respectively, and both are transition states characterized by a negative curvature direction with imaginary frequencies of 30i cm^{-1} for the *trans* and 77i cm^{-1} for the *cis* conformers. Inclusion of correlation corrections has an interesting effect. The ordering of energies for the structures reverses between $\tau = 123$ and 133° so that the barrier disappears. This suggests that there probably is no *trans*-skew minimum. The energies of these two structures also become closer to the *cis*-skew energy. In contrast, correlation raises the energy of the *cis* and *trans* transition states by about 0.8 kcal/mol. Thus the curvature about the *cis*-skew minimum broadens toward larger τ and the curve only becomes steeper for $\tau > 133°$. The curvature towards the *cis* side becomes somewhat steeper. The correlated result for the barrier height between the *cis*- and the *trans*-skew conformers (2.5 kcal/mol) is now in very good agreement with the experimental value of 2.8 ± 0.4 kcal/mol. (The calculated effect of zero-point energy differences is to lower the barrier heights by only about 0.1 kcal/mol.)

Our model for the preference of the *cis*-skew structure over the *trans* is based primarily on minimizing the repulsive 1,3 C–F bond dipole interactions and, secondarily, by minimizing the steric interactions of the fluorines in the 1,3 positions. In order to obtain more information about such effects, we investigated a model for the polymer derived from the head-to-tail polymerization of fluoroacetylene, 1,3-difluoro-1,3-butadiene. Since there is only one 1,3 interaction present in this compound, the *trans* conformer could be a mimimum energy structure, and our calculations indeed confirm this is the case (Table II). Surprisingly, however, an almost planar *cis* structure ($\tau = 7°$) also is calculated to be an energy minimum structure, and it actually is lower in energy than the *trans* structure by 0.6 kcal/mol (0.4 kcal/mol at the SCF level)! We surmise the *cis* conformation is stabilized by an internal hydrogen bond between the terminal CH and CF

Figure 2. CPK model of perfluoro(1,3)butadiene. Image generated on a CRAY X-MP supercomputer using OASIS software. The carbons are the dark atoms and the fluorines are the light atoms.

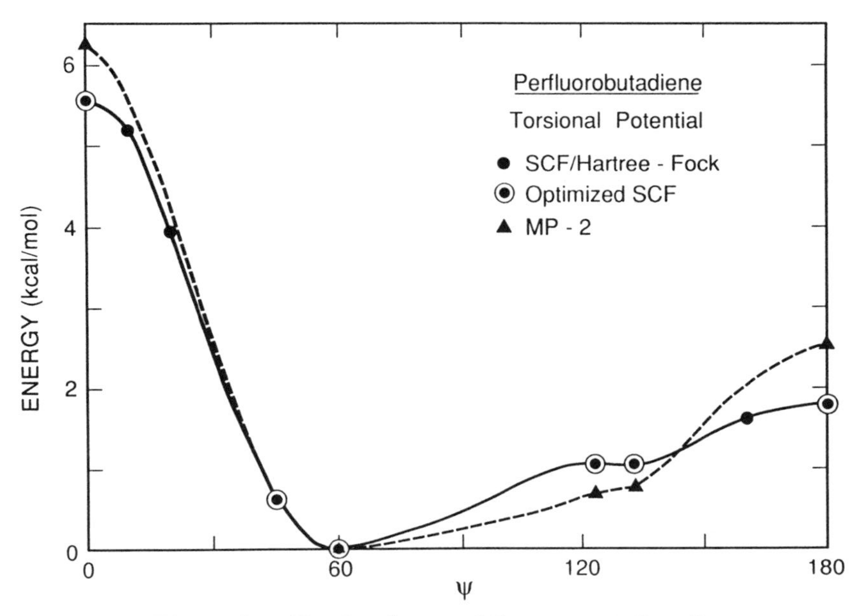

Figure 3. Torsional potential energy surface for perfluoro(1,3)butadiene.

groups. The non-bonded H···F distance is 2.40 Å, which is 0.37 Å less than the sum of H and F van der Waals radii, and compares to a CF···HO hydrogen bond length of 2.37 Å in 2-fluoroacetaldehyde enol where the hydrogen bond strength is 3.2 kcal/mol (see below) (*26*).

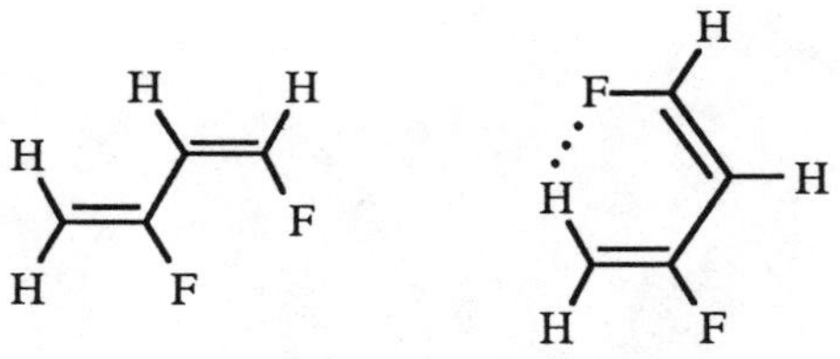

Trans *Cis*

Table II. Polyfluoracetylene Model Structures

CHF=CHCF=CH2	*ΔE(SCF)*[a]	*ΔE(MP-2)*[a]
Trans	0.42	0.59
Cis (τ= 7°)	0.00	0.00
CHF=CHCH=CFCF=CH2		
Trans-HH	0.0	0.0
CHF=CHCF=CHCF=CH2		
Cis-HT	4.5	3.3
Trans-HT	5.4	4.5

[a]Relative energies in kcal/mol.

1,3,5-Hexatrienes. To gain further insight into the conformation of fluorinated polyacetylene chains, we also calculated the structures of some fluorinated hexatrienes. The relative energetics of various conformers of perfluorohexatriene are given in Table III and four important conformers are shown in Figure 4. The various conformers can be related to the structures of perfluorobutadiene. In perfluorohexatriene, there are two low energy *cis*-skew structures – one with the two torsions in the same direction and the other with the torsions in opposite directions. The skew angles are 5-10° smaller than the angle in perfluorobutadiene and are approximately equal. The energies of these two structures are nearly identical. Two other minima corresponding to *trans*-skew structures were found at the SCF level.

In this case, the structure whose torsion angles are of opposite sign is higher in energy. The structures are 2.0 and 2.3 kcal/mol above the *cis*-skew minima. In contrast to the results for the butadiene, correlation raises the energies. The all-*trans* structure is no longer a transition state as in perfluorobutadiene but is a saddlepoint characterized by two imaginary frequencies. As expected, there is a significant correlation effect to the barrier, which is essentially twice that calculated for the butadiene. The lowest torsion frequency is the symmetric motion for the conformers with the same sign in the torsions and asymmetric for those of opposite sign, and is about the same as that found in the butadiene. The other torsion frequency is about 50% higher, but we note that the torsion frequencies are all still quite low and the molecule is extremely flexible toward twisting about the C–C single bonds. The all-*trans* conformer shows steric interactions somewhat larger than those in butadiene. The 1,3 F···F interactions are 2.66 Å in the butadiene, but decrease to 2.61 and 2.58 Å for the 1,3 and 3,5 interactions, respectively, in the hexatrienes. Because of the additional double bond, there is not as much flexibility in the bending angle distortions and the 1,3-nonbonded distances are of necessity shorter.

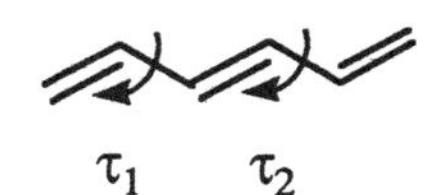

Table III. Relative Energies of Perfluoro-1,3,5-hexatriene Conformers

τ_1 [a]	τ_2 [a]	$\Delta E(SCF)$ [b]	$\Delta E(MP\text{-}2)$ [b]	ν_1 [c]	ν_2 [c]
180	180	3.10	4.42	26i	39i
146	146	2.01	2.22	24	33
146	-146	2.29	2.46	18	41
52	52	0.00	0.00	36	70
53	-53	0.10	-0.16	40	77

[a]Torsion angles in degrees. [b]Relative energies in kcal/mol. [c]Torsional frequencies in cm^{-1}.

These model studies show that we cannot distinguish at this level if there is a preferential twist in poly(difluoropolyacetylene). However, we can definitively conclude that it will not be planar. The reported crystal structure of $C_6H_5(CF)_8C_6H_5$ (*37*) confirms this expectation. The molecule sits on a center of symmetry and the torsion about the central C(F)–C(F) bond is 180°C. The other two C(F)–C(F) torsions are 47° and are twisted in opposite directions. Thus if the perfluoro polymer can be synthesized, its

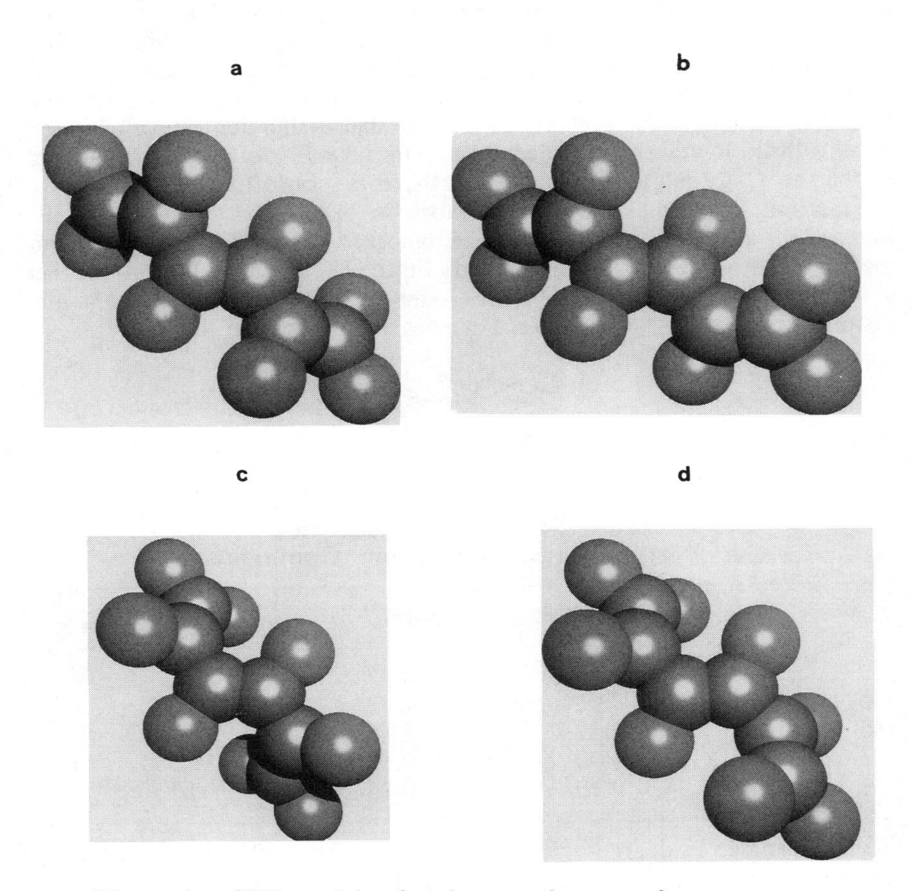

Figure 4. CPK models of various conformers of perfluoro(1,3,5)hexatriene. (a) $\tau_1 = 146$, $\tau_2 = 146$; (b) $\tau_1 = 146$, $\tau_2 = -146$; (c) $\tau_1 = 52$, $\tau_2 = 52$; (d) $\tau_1 = 53$, $\tau_2 = -53$.

properties should differ significantly from those of the parent hydrocarbon. Of course, polyacetylene needs to be doped to be conductive. Doping the perfluoro compound, however, likely will lead to significantly different results. For example, work on fluorinated allyl anions has shown that these ions prefer non-planar structures (*38*). Thus addition of an electron may not make the system planar. Oxidation has a better chance of inducing planarity, but removal of an electron from the chain could be difficult since fluorine is a strong electron withdrawing substituent, and the cation will be destabilized (*39*).

We also examined trifluoro hexatriene models for poly(fluoracetylene), both the head-to-tail (HT) and head-to-head (HH) polymers. The energetic results are shown in Table II and the structures are shown schematically below. The HH isomer is more stable than the HT isomer. Of even more interest, the *trans*-HT isomer is predicted to be 1.2 kcal/mol less stable than the *cis* isomer. This is consistent with the butadiene results and the effect is essentially additive. These differences can be explained by examining the number of internal hydrogen bonds and repulsive fluorine interactions. The HH isomer for hexatriene has four possible hydrogen bonds, two with lengths of 2.46Å, and two of 2.56 Å. In the *cis*-HT isomer there are two H···F hydrogen bonds (2.32 Å) and the least stable *trans*-HT isomer has no hydrogen bonds and unfavorable dipolar interactions between the C–F bonds (and between the C–H bonds).

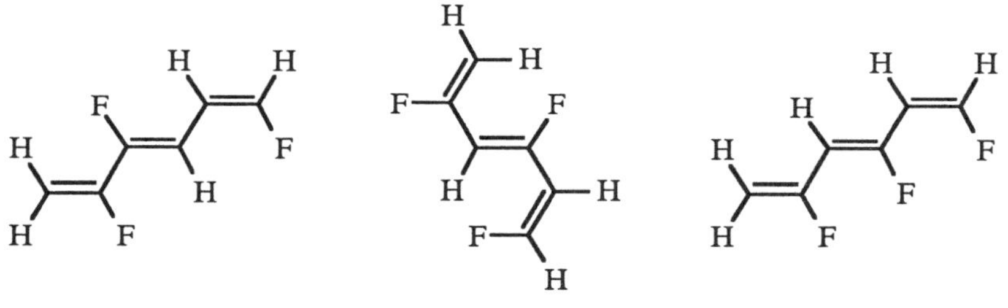

Trans-HH *Cis*-HT *Trans*-HT

These results suggest that poly(fluoroacetylene) will have quite different properties depending on how fluoroacetylene polymerizes. The head-to-head polymer should have an all *trans* structure and could form an extended hydrogen-bonded network to form sheets as shown below. The network is formed by slipping one chain with respect to the other. For the *cis*, *cis* head-to-tail isomer a similar network can form with chains simply reversed. In both cases, the dipole of the sheet should be small. For the *trans* conformer of the head-to-tail polymer, a hydrogen bonding network also can be set up, but in this case there will be a large net dipole in the plane. This behavior is reminiscent of poled poly(vinylidinefluoride) (*40, 41*) and suggests that if the polymer can be made it could have unique ferroelectric or piezoelectric properties. Furthermore, because it is a polyene, it should have a significantly lower band gap relative to the polyalkane and could have unique electrical properties.

The electronic structures of poly(fluoroacetylene) and poly(difluoroacetylene) have been investigated previously using the *ab initio* Hartree-Fock crystal orbital method with a minimum basis set (*42*). Only the *cis* and *trans* isomers with assumed, planar geometries were studied. The *trans* isomer was calculated to be more stable in both cases, and the *trans* compounds were predicted to be better intrinsic semiconductors and more conductive upon reductive doping than *trans* polyacetylene. However, our results show that head-to-tail poly(fluoroacetylene) prefers the *cis* structure and that the *trans* structure for poly(difluoroacetylene) will not be stable. Thus the conclusions reached previously need to be re-evaluated based on our new structural information. Furthermore, as noted above, addition of electrons to these polymers may lead to structural deformations that could significantly change the conductive nature of the materials.

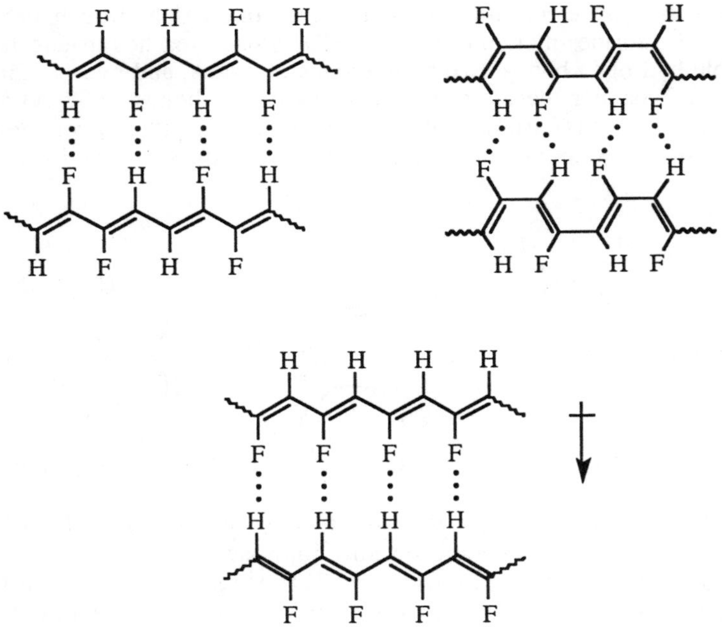

F H F H
H F H F
F H F H
H F H F

F H F H
H F H F
F H F H
H F H F

H H H H
F F F F
H H H H
F F F F

H···F Hydrogen Bonding in Enols and Alcohols. As noted above, there are important hydrogen bonding (CF···HC) contributions to the stability of the model poly(fluoroacetylenes). We also have investigated the nature of intramolecular CF···HO hydrogen bonding in the model compounds 2-fluoroethanol and 2-fluoroacetaldehyde enol. This study was prompted by reports on the structure and stability of 2-fluoroethanol (*43*) and the recent synthetic work on fluoroenolates (*44*).

The most stable structure for 2-fluoroethanol is the hydrogen-bonded structure shown below. Although the GG conformer is estimated experimentally to be more stable than TT by at least 2 kcal/mol, it was not possible to determine the strength of the internal hydrogen bond. The computed energetics are given in Table IV. The ΔH values are corrected for zero-point effects. The energy difference between the GG and GT structures is 1.9 kcal/mol and corresponds to the strength of the internal hydrogen bond. The difference in the GG and GT structures is only 0.1 kcal/mol and corresponds to the gauche effect, which is smaller than the effect in 1,2-difluoroethane where the gauche is more stable than the *trans* conformer by 0.8 kcal/mol (*45, 46*). For direct comparison with experiment, we calculated the S° values from our optimized structures and calculated frequencies. The calculations give ΔG°(TT-GG)=ΔG°(GT-GG)=1.7 kcal/mol. Our calculated value is in excellent agreement with the difference in energy derived from band intensities in CCl_4 solution (*47*) which gives 2.07 ± 0.53 kcal/mol for the stabilization of the hydrogen-bonded form. Other experimental values range from the estimate of 2.7 (+1.8, -1.5) kcal/mol from electron diffraction measurements (*43*) to 1.05 from NMR data (*48*).

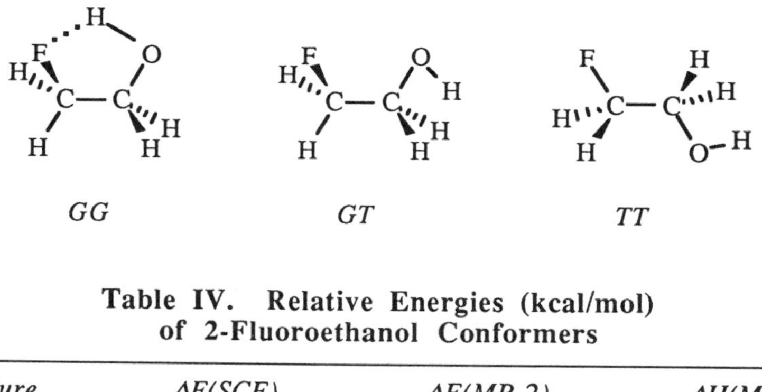

GG *GT* *TT*

Table IV. Relative Energies (kcal/mol) of 2-Fluoroethanol Conformers

Structure	*ΔE(SCF)*	*ΔE(MP-2)*	*ΔH(MP-2)*
GG	0.00	0.00	0.00
GT	1.90	2.12	1.93
TT	1.50	2.25	2.05

The nature of the hydrogen bond in 2-fluoroacetaldehyde enol is quite different from that in the alcohol. Among the four structures shown below, the *cis-syn* structure is the most stable (Table V). The difference in energies of the *cis-syn* and *cis-anti* structures gives an estimate of the hydrogen bond energy. At the MP-2 level, this is 3.53 kcal/mol and corrections for zero-point effects yield a value of 3.21 kcal/mol. (The zero-point effects are significantly larger in the enol than in the alcohol.) The difference in energy between the *cis-syn* and *trans-syn* structures is 4.45 kcal/mol at the MP-2 level. Subtracting the value for the hydrogen bond strength gives 0.92 kcal/mol (0.87 kcal/mol when corrected for zero-point effects), which is nearly identical to the calculated *cis-trans* energy difference of 0.96-1.00 kcal/mol for 1,2-difluoroethylene (*49*) and the experimental energy difference of 1.08 kcal/mol (*50-52*). The difference between the *cis-anti* and *trans-anti* structures is even larger, 1.43 kcal/mol. Thus the *cis* effect in the ethylene is not dependent on whether OH is substituted for F, whereas in the ethane this clearly makes a large difference.

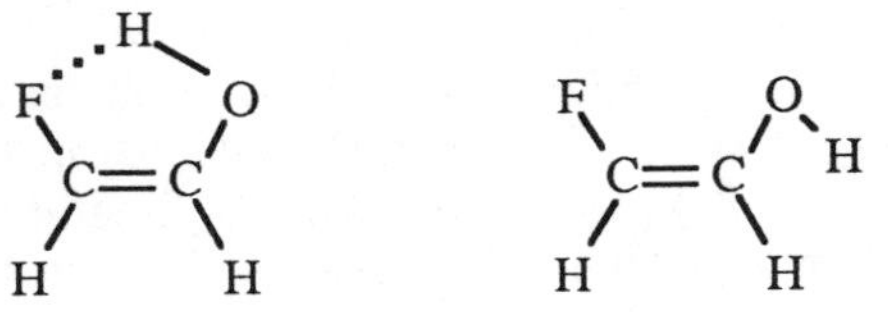

Cis-syn *Cis-anti*

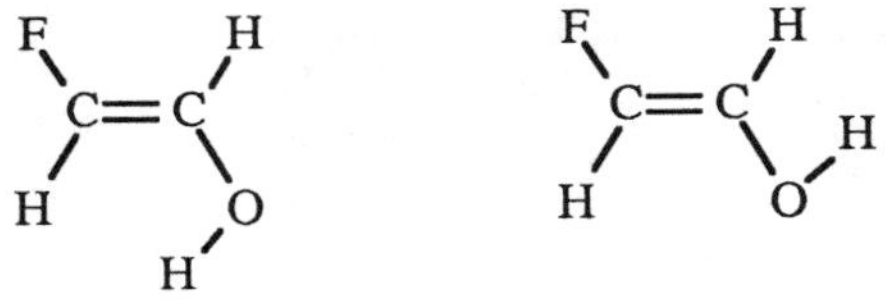

Trans-syn *Trans-anti*

Table V. Relative Energies (kcal/mol) of 2-Fluoroacetaldehyde Enol Structures

Structure	*ΔE(SCF)*	*ΔE(MP-2)*	*ΔH(MP-2)*
Cis-syn	0.00	0.00	0.00
Cis-anti	3.30	3.53	3.21
Trans-syn	4.13	4.45	4.08
Trans-anti	4.35	4.96	4.40

The strength of the hydrogen bond in the enol is 1.7 times the hydrogen bond strength of the alcohol (3.2 vs. 1.9 kcal/mol). This is probably due to the better planar five-membered ring geometry for intramolecular bonding that can be attained in the enol. The OH···F bond length in the enol is 2.37 Å compared with 2.52 Å in the alcohol. Both are shorter than the sum of the van der Waals radii of H and F, but the enol clearly has the stronger interaction. These differences are also reflected in the OH stretching frequency. The calculated OH stretch in the GG structure of the alcohol is 18 cm^{-1} below the value in the GT structure. In the enol, a greater difference of 56 cm^{-1} is predicted between the *cis-syn* and *cis-anti* structures, which again is consistent with the stronger hydrogen bond in the enol.

Alkali Ion Bridging in Enolates. It is clear that hydrogen bonding further stabilizes the *cis* enol. Because of their synthetic importance, it is of interest to examine the ability of alkali ions to stabilize enolates. Can the alkali metal cation also strongly interact with an adjacent fluorine, potentially forming a bridge? Experimental structure data indeed suggest the C-F bond can coordinate to metal cations, but little is known about the interaction energies (*53, 54*). We therefore computed the energies of structures wherein the alkalis Li, Na, and K were substituted for H in the enol. The alkalis clearly interact strongly with the F in the *cis* isomer. The *trans-cis* energy difference for the free enolate is 1.6 kcal/mol at the SCF level and 1.8 kcal/mol at the MP-2 level (Table VI), almost double the energy difference in 1,2-difluoroethylene. As noted above, the addition of H^+ stabilizes the *cis* isomer by 4.45 kcal/mol over the *trans* at the MP-2 level. The addition of Li^+ stabilizes the *cis* isomer by a much larger amount, 13.2 kcal/mol. Similarly, Na^+ stabilizes the *cis* isomer by 13.6 kcal/mol and K^+ by 10.8 kcal/mol. These results suggest a very strong interaction of the alkali metal cation with O and F. The bond lengths confirm this (Table VI). The O-M bond distance lengthens by 0.13 Å in the *cis* compared to the *trans* isomer for all three alkalis. The M-F bond distance for Li and Na is only 0.11 Å longer than the M-O distance and for K the difference is only 0.15 Å. This demonstrates that there are very strong interactions between the M and F and these are comparable to the M and O interactions.

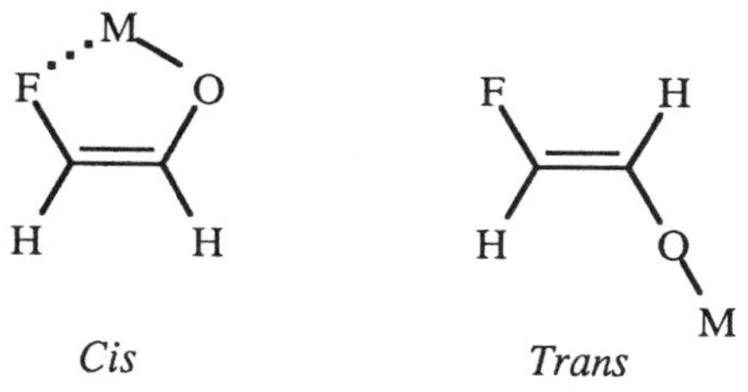

Cis *Trans*

Table VI. Relative Energies and Bond Lengths of 2-Fluoroacetaldehyde Enolate Isomers

	ΔE (Trans-cis)[b]		*R(M-F)*[c]	*R(M-O)*[c]	
M	*SCF*	*MP-2*	*Cis*	*Cis*	*Trans*
- [a]	1.6	1.8	-	-	-
Li	13.2	13.2	1.86	1.75	1.62
Na	12.4	13.6	2.21	2.10	1.97
K	10.3	10.8	2.62	2.47	2.34

[a]Free anion. [b]Energy difference in kcal/mol. [c]Bond lengths in Å.

Conclusions

The above results again demonstrate that fluorination can have important effects on the molecular structures of organic systems. The effect in polyacetylene is to dramatically change its structure which in turn should alter its electronic properties. For the perfluorinated polymer, a *cis*-skew structure will be adopted whereas head-to-tail polymerization of fluoroacetylene will lead to a *cis* structure. This latter conformation is stabilized by an internal CH···F hydrogen bond. In a similar fashion, good internal OH···F hydrogen bonds are formed in 2-fluoroethanol and 2-fluoroacetaldehyde enol. Substitution of an alkali for H in the enol leads to a signifcantly stronger interaction where the metal cation strongly interacts with both the O and the F to dramatically stabilize the *cis* enolate.

The calculations discussed in this chapter represent just a fraction of the theoretical work being done on fluorinated systems. The computational results have significantly enhanced our understanding of these novel materials and provide unique insights into molecular behavior. Further work will continue in this field with the mainstay being traditional *ab initio* molecular orbital theory. However, applications of new methods to fluorinated systems are now under investigation. The most important of these could be the local density functional method (*55-58*). The computational advantage of this method is that it scales as N^3 where N is the number of basis functions, compared with the N^4 scaling of the Hartree-Fock method. Thus, much larger systems can be treated with the inclusion of correlation corrections in even the geometry optimizations. Initial studies of the method for fluorinated systems are very promising and even more areas of fluorine chemistry may be opened to theoretical study (*14*).

Literature Cited

1. Smart, B. E. In *Molecular Structure and Energetics*; Liebman, J. F.; Greenberg, A., Eds.; VCH Publishers: Deerfield Beach, FL, 1986, Vol. 3, Chapter 4.
2. Smart, B.E. In *The Chemistry of Functional Groups, Supplement D*, Part 2; Patai, S.; Rappaport, Z., Eds., John Wiley & Sons: New York, 1983, Chapter 14.
3. Förster, H.; Vögtle, F. *Angew. Chem. Int. Ed. Engl.* **1977**, *16*, 429.
4. Bondi, A. *J. Phys. Chem.* **1964**, *68*, 441.
5. Dixon, D. A.; Smart, B. E.; Fukunaga, T. *J. Am. Chem. Soc.* **1986**, *108*, 1585.
6. Hehre, W. J.; Radom, L.; Schleyer, P. vR.; Pople, J. A. *Ab Initio Molecular Orbital Theory*; John Wiley & Sons: New York, 1986.
7. Schaefer, H. F. *The Electronic Structure of Atoms and Molecules*; Addison-Wesley: Reading, MA, 1972.
8. Dixon, D. A. *J. Phys. Chem.* **1988**, *92*, 86.
9. Dixon, D. A.; Smart, B. E.; Fukunaga, T. *J. Am. Chem. Soc.* **1986**, *108*, 4027.
10. Dixon, D. A. *J. Phys. Chem.* **1986**, *90*, 2038.
11. Dixon, D. A.; Van-Catledge, F. A. *Int. J. Supercomputer Appli.* **1988**, *2*, (2), 62.
12. Dixon, D. A. In *Science and Engineering on Cray Supercomputers*, Proceedings of the Third International Symposium, Cray Research, Minneapolis, MN, 1987, p. 169.
13. Dixon, D. A.; Capobianco, P. J.; Mertz, J. E.; Wimmer, E. In *Science and Engineering on Cray Supercomputers*, Proceedings of the Fourth International Symposium, Cray Research, Minneapolis, MN, 1988, p. 189.
14. Dixon, D. A.; Andzelm, J.; Fitzgerald, G.; Wimmer, E.; Delley, B. In *Science and Engineering on Cray Supercomputers*, Proceedings of the Fifth International Symposium, Cray Research, Minneapolis, MN, 1990.
15. Chien, J. C. W. *Polyacetylene: Chemistry, Physics and Materials Science*; Academic Press: New York, 1984.
16. GRADSCF is an *ab initio* program system designed and written by A. Komornicki at Polyatomics Research.
17. Komornicki, A.; Ishida, K.; Morokuma, K.; Ditchfield, R.; Conrad, M. *Chem. Phys. Let.* **1977**, *45*, 595.
18. McIver, J. W., Jr.; Komornicki, A. *Chem. Phys. Lett.* **1971**, *10*, 202.
19. Pulay, P. In *Applications of Electronic Structure Theory*; Schaefer, H. F. III, Ed.; Plenum Press: New York, 1977, p.153.
20. King, H. F.; Komornicki, A. *J. Chem. Phys.* **1986**, *84*, 5465.
21. King, H. F.; Komornicki, A. In *Geometrical Derivatives of Energy Surfaces and Molecular Properties*; Jorgenson, P.; Simons, J., Eds.; NATO ASI Series C; D. Reidel: Dordrecht, 1986, Vol. 166, p. 207.

22. Möller, C.; Plesset, M. S. *Phys. Rev.* **1934**, *46*, 618.
23. Pople, J. A.; Binkley, J. S.; Seeger, R. *Int. J. Quantum Chem. Symp.* **1976**, *10*, 1.
24. Dunning, T. H., Jr.; Hay, P. J. In *Methods of Electronic Structure Theory*; Schaefer, H.F., III, Ed.; Plenum Press: New York, 1977, Chapter 1.
25. Dunning, T. H. *J. Chem. Phys.* **1971**, *55*, 716.
26. Dixon, D. A.; Smart, B. E. *J. Phys. Chem.*, submitted 1990.
27. Dixon, D. A.; Gole, J. L.; Komornicki, A. *J. Chem. Phys.* **1988**, *92*, 1378.
28. Partridge, H.; Dixon, D. A.; Walch, S. P.; Bauschlicher, C. W., Jr.; Gole, J. L. *J. Chem. Phys.* **1983**, *79*, 1859.
29. Almenningen, A.; Traetteburg, M. *Acta Chem. Scand.* **1958**, *12*, 1221.
30. Alberts, I. L.; Schaefer, H. F. III *Chem. Phys. Lett.* **1989**, *161*, 375.
31. Rice, J. E.; Liu, B.; Lee, T. J.; Rohlfing, C. M. *Chem. Phys. Lett.* **1989**, *161*, 277.
32. Bock, C. W.; George, P.; Trachtman, M. *Theor. Chim. Acta* **1984**, *64*, 293.
33. Carriera, L. A. *J. Chem. Phys.* **1975**, *62*, 3851.
34. Chang, C. H.; Andreassen, A. L.; Bauer, S. H. *J. Org. Chem.* **1971**, *36*, 920.
35. Brundle, C. R.; Robin, M. *J. Am. Chem. Soc.* **1970**, *92*, 5550.
36. Lorig, G., OASIS program system, Cray Research, Inc., Eagan, MN, 1988.
37. Yurchenko, V. M.; Antipin, M. Yu.; Struchkov, Yu. T.; Yagupolski, L. M. *Cryst. Struct. Comm.* **1978**, *7*, 77.
38. Dixon, D. A.; Smart, B. E.; Fukunaga, T. *J. Phys. Org. Chem.* **1988**, *1*, 153.
39. Dixon, D. A.; Eades, R. A.; Frey, R.; Gassman, P. G.; Hendewerk, M. L.; Paddon-Row, M. N.; Houk, K. N. *J. Am. Chem. Soc.* **1984**, *106*, 3885.
40. Lovenger, A. J. *Science* **1983**, *220*, 115.
41. Lovenger, A. J. In *Developments in Crystalline Polymers*; Gassett, D. C., Ed.; Applied Sciences Publishers: London, 1982, Chapter 5.
42. Bakhshi, A. K.; Ladik, J.; Liegener, C.-M. *Synth. Metals* **1987**, *20*, 43.
43. Huang, J.; Hedberg, K. *J.Am. Chem. Soc.* **1989**, *111*, 6909.
44. Qian, C.-P.; Nakai, T.; Smart, B. E.; Dixon, D. A. *J. Am. Chem. Soc.* **1990**, *112*, 4602.
45. Dixon, D. A.; Smart, B. E. *J. Phys. Chem.* **1988**, *92*, 2729.
46. Hirano, T.; Nonoyama, S.; Miyajima, T.; Kurita, Y.; Kawamura, T.; Sato, H. *J. Chem. Soc. Chem. Commun.* **1986**, 606.
47. Kruger, P. J.; Mettee, H. D. *Can. J. Chem.* **1964**, *82*, 326.
48. Pachler, K. G. R.; Wessels, P. L. *J. Mol. Structure* **1970**, *6*, 471.

49. Dixon, D. A.; Smart, B. E.; Fukunaga, T. *Chem. Phys. Lett.* **1986,** *125,* 447.
50. Craig, N. C.; Entemann, E. A. *J. Am. Chem. Soc.* **1961,** *83*, 3047.
51. Craig, N. C.; Overend J. *J. Chem. Phys.* **1969,** *51*, 1127.
52. Craig, N. C.; Piper, L. G.; Wheeler, V. L. *J. Phys. Chem.* **1971,** *75*, 1453.
53. Murray-Rust, P.; Stallings, W. C.; Monti, C. T.; Preston, R. K.; Glusker, J. P. *J. Am. Chem. Soc.* **1983,** *105*, 3206.
54. Carrell, H. Li; Glusker, J. P.; Piercy, E. A.; Stallings, W. C.; Zacharias, D. E.; Davis, R. L.; Astbury, C.; Kennard, C. H. L. *J. Am. Chem. Soc.* **1987,** *109*, 8067.
55. Parr, R. G.; Yang, W. *Density Functional Theory of Atoms and Molecules*; Oxford University Press: New York, 1989.
56. Salahub, D.R. In *Ab Initio Methods in Quantum Methods in Quantum Chemistry-II*; Lawley, K. P., Ed.; J. Wiley & Sons: New York, 1987, p. 447.
57. Wimmer, E.; Freeman, A. J.; Fu, C.-L.; Cao, P.-L.; Chou, S.-H.; Delley, B. In *Supercomputer Research in Chemistry and Chemical Engineering*; Jensen, K.F.; Truhlar, D. G., Eds.; ACS Symposium Series; American Chemical Society: Washington, D.C., 1987, p. 49.
58. Delley, B. *J. Chem. Phys.* **1990,** *92*, 508.

RECEIVED October 19, 1990

Chapter 3

Systematics and Surprises in Bond Energies of Fluorinated Reactive Intermediates

Joel F. Liebman[1], Sharon O. Yee[1], and Carol A. Deakyne[2,3]

[1]Department of Chemistry and Biochemistry, University of Maryland, Baltimore County Campus, Baltimore, MD 21228
[2]Ionospheric Interaction Division, Air Force Geophysics Laboratory, Hanscom Air Force Base, MA 07131

Despite the fact that fluorine has the highest ionization potential and electronegativity of all the "chemical" elements, positively charged fluorinated species enjoy a rich and diverse organic chemistry. The general observations that fluorine has little effect on the ionization potential of planar molecules (the "per" and π-fluoro effects) and that bond dissociation energies are less dependent on substitution when the species of interest and its fragments correlate (have the same number of σ and π electrons) provide powerful insights into the energetics of fluorinated reactive intermediates.

Positively Charged Fluorine-Containing Species and the Estimation of ionization potentials.

It is well-established that fluorine has the highest ionization potential and electronegativity (1) of the elements that are known to form isolable compounds. We thus ignore helium and neon (2) even though ions such as He_2^+ and NeH^+ enjoy gas phase stability comparable to many "normal" species -- "light" noble gas cations are seemingly too oxidizing and/or acidic for solvent, other solutes or salt to survive. Relatedly, although most assuredly F^+ exists as a stable species in the gas phase (3), F^+ is at most a formal construct to the solution chemist. It is not **even** a reactive intermediate -- unlike its congeners (the "other" monoatomic halogen cations Cl^+,Br^+,I^+) F^+ has seemingly never been

[3]Current address: Department of Chemistry, Eastern Illinois University, Charleston, IL 61920

0097–6156/91/0456–0036$06.00/0

observed in either solvated or crystalline form. Nonetheless, a rich cation chemistry is enjoyed by fluorinated organic compounds. (See, for example, ref. 4, wherein both fluoronium ions and fluorocarbonium ions are discussed.) This is largely a consequence of the fact that the ionization potentials of fluorinated compounds do not reflect the high ionization potential of the component fluorine atoms. Indeed, from our reference 3, the compendium of by Lias and her coworkers for the gas phase energetics of neutral and ionic species, it is seen that molecular fluorine has an ionization potential of 15.698 eV ($\cong$ 362 kcal mol^{-1}), significantly lower than that of atomic fluorine, IP(F) = 17.422 eV ($\cong$ 402 kcal mol^{-1}). The ionization potential of molecular fluorine is also lower than that for hydrogen fluoride, 16.044 eV ($\cong$ 370 kcal mol^{-1}).

If we view fluorine as generally electron withdrawing and hydrogen, by contrast, as electron donating, the near equality of the ionization potentials of the two diatomic molecules is surprising. However, the near equality of IP(F_2) and IP(HF) is not fortuitous. We recall the "perfluoro effect" (5,6) that asserts the energy for the ionization of a π-electron from a planar, perfluorinated, species is comparable to that of the parent "hydrogenated" species while the energy for the ionization of a σ-electron is considerably higher for the former fluorinated species. We also recall its generalization, the "π-fluoro effect" (7), that asserts the π-ionization potential of an arbitrarily partially fluorinated species is nearly the same as its all-fluorine and all-hydrogen counterparts. Indeed, one could even conclude that fluorination tends to decrease ionization potentials of free radicals. In contrast, for nonplanar species where σ- and π-orbitals are by definition quantum mechanically inseparable, total or partial replacement of hydrogen by fluorine generally increases the ionization potentials. In the text that follows, we will always use the term "ionization potential" for the process of removing the least bound electron of the molecule and letting the geometry "relax," in particular a π-electron should the molecule be planar -- this is in lieu of the more formally correct term "adiabatic ionization potential." We thus note that ecause the diatomic molecules, F_2 and HF, are linear they are also by definition planar. Their ionization potentials are thus expected to be of "similar" magnitude.

To document the generality of the small changes in π-ionization potentials upon fluorination, the reader is referred to Table I. Here are found all of the ionization potentials for benzene and all 12 of its partially fluorinated derivatives, and the perfluorinated hexafluorobenzene. (All of these numbers are in eV, where 1 eV $\cong$ 23.06 kcal mol^{-1}.)

"Similar" in this current case of benzene and its

Table I. Ionization Potentials (in eV) of Benzene and its Fluorinated Derivatives

n_F	(posns.)	IP	(posns.)	IP	(posns.)	IP
0		9.25				
1		9.20				
2	(1,2)	9.28	(1,3)	9.33	(1,4)	9.19
3	(1,2,3)	9.7	(1,2,4)	9.30	(1,3,5)	9.67
4	(1,2,3,4)	9.53	(1,2,3,5)	9.53	(1,2,4,5)	9.35
5		9.63				
6		9.91				

partially and totally fluorinated derivatives corresponds to ca. 0.7 eV or 16 kcal mol^{-1}. The reader may protest that this range of ionization potentials is so large as to establish the lack of "similarity" previously proposed. However, we note that the ionization potentials do not uniformly increase with increasing number of fluorines. It is also not even particularly more constant within a set of isomers. Within the current context and definition of "similar," the perfluoro and π-fluoro effects may be used to assist and to extend our understanding of the energetics of fluorinated species. In particular, several important classes of fluorinated organic compounds and related reactive intermediates are examined. However, in the name of brevity, we will limit our attention to those species that contain two carbons and to those one-carbon compounds where particularly salient results arise, and leave the study of "larger" species to future investigation by the readers and authors alike.

C-C Bond Strengths in Ethane and Hexafluoroethane and in their Radical Cations

We now turn to the energetics of 2-carbon species and commence with the saturated ethane derivatives. As a further desire for brevity, we will only consider the parent ethane and hexafluoroethane. The C-C bond strengths in ethane and hexafluoroethane are taken here to be the enthalpy of the homolysis reactions (equations 1 wherein X = H and F respectively, where the heats of formation, $\Delta H_f°(g)$, of the ethane and of the methyl radicals are from ref. 3).

$$C_2X_6 \longrightarrow 2\ CX_3\cdot \qquad (1)$$

We thus find D(C-C, C_2H_6) = 90 kcal mol^{-1} and D(C-C, C_2F_6) = 101 kcal mol^{-1}. Is it reasonable that D(C-C, C_2H_6) < D(C-C, C_2F_6) ? The following are some of the differences of the parent and perfluorinated ethanes chronicled by Smart (8) and their impact on their corresponding C-C bond strengths. Simple electronegativity reasoning suggests that the carbons are more positive in hexafluoroethane than in ethane. Therefore, electrostatics implies that the fluorinated species would have the weaker C-C bond. However, this same electronegativity reasoning, accompanied by Bent's rule (electronegative atoms prefer bonding to p orbitals), suggests that the C-C bond in hexafluoroethane will have high s character. As such, it is expected to be anomalously strong. There are also vicinal substituent effects: the fluorines on the "opposite carbons" attract these carbons and repel their fluorines. What do we now conclude about the C-C bond in C_2F_6? Is it strong or weak or even normal? We note that to the consternation of qualitatively minded theorists, e.g. ref. 9, the hydrocarbon free radical $CH_3\cdot$ is planar while the fluorinated radicals $CH_2F\cdot$, $CHF_2\cdot$ and $CF_3\cdot$ are increasingly pyramidal (10). Quite fortuitously, the F-C-F angle in $CF_3\cdot$ is found (10) to have nearly the idealized tetrahedral angle. Should one modify the geometry of the $CH_3\cdot$ to mimic the $CF_3\cdot$ (using the quantum chemically calculated "tetrahedralization energy" of ca. 7 kcal mol^{-1} (11), it will become suitable for direct dimerization to the ethane without either geometric or electronic reorganization. So doing, the "new" bond strength of C_2H_6 is seemingly larger than that of C_2F_6. (The reader may note a related analysis of the C-C bond strengths in ethane, neopentane, 2,2,3,3-tetramethylbutane and diamond as part of a discussion on the tetravalence of carbon (12).)

In a related way, one can define the C-C bond strengths of the radical cation of the parent and perfluorinated ethanes by the use of equation (2)

$$C_2X_6^+ \longrightarrow CX_3\cdot + CX_3^+ \quad (2)$$

Our intuition derived from the per and π-fluoro effects suggests that fluorination of the nonplanar ethane should be accompanied by an increased ionization potential. Reference 3 corroborates this: C_2H_6 and C_2F_6 have ionization potentials of 11.52 and 13.4 eV respectively. Likewise, our intuition suggests that the ionization energies of planar $CH_3\cdot$ and planar $CF_3\cdot$ should be comparable. From the definition of the inversion barrier, ΔE_{inv}, as the energy difference of the planar and pyramidal forms and the identity for adiabatic ionization potentials (equation 3),

$$IP_{pyramidal}(CF_3\cdot) = IP_{planar}(CF_3\cdot) + \Delta E_{inv}(CF_3\cdot) \quad (3)$$

we would conclude that the ionization potential of pyramidal $CF_3\cdot$ is greater than that of planar $CH_3\cdot$, 9.81 eV. In fact, while the ionization potential of $CF_3\cdot$ is still contested (13-15), it is unequivocally less than 8.9 eV. This suggests there is considerable stabilization of CF_3^+, a finding predicted by the importance of the three equivalent doubly bonded resonance structures $F^+{=}C(F)_2$ that accompany the electron-deficient $C^+(F)_3$. This stabilization also parallels the recent analyses (16, 17) of the resonance energy in the isoelectronic BF_3 that gave a value of ca. 40 kcal mol^{-1}. Intuitively, resonance energy is maximized when the resonance structures are equivalent. For example, the charge delocalization in a complex formed between a neutral and its derived cation should be maximized when the two halves have the same geometry. In the current case, they do not: we recall that $CF_3\cdot$ is pyramidal while CF_3^+ is planar. It is thus not altogether surprising that the bonding of CF_3^+ with CF_3 is weaker than CH_3^+ with CH_3. Equivalently, the C-C bond strength in $C_2F_6^+$ is expected to be less than in $C_2H_6^+$. What is most surprising is that from the numbers above, we derive D(C-C, $C_2H_6^+$) = 50 kcal mol^{-1} and D(C-C, $C_2F_6^+$) = -3 kcal mol^{-1}. That is, while ethane radical cation has a bond strength about one half of that of the neutral hydrocarbon as befits a one-electron bond, the radical cation of hexafluoroethane is seemingly unbound! (See ref. 18 for a discussion of the role of electronic state on the fragmentation of $C_2F_6^+$.)

What is the C=C bond strength in C_2H_4 and C_2F_4?

By analogy to other multiply bonded species such as O_2, dissociation of multiply bonded species can be defined by equation 4.

$$D(C_2X_4) = 2\Delta H_f^o(CX_2) - \Delta H_f^o(C_2X_4) \tag{4}$$

Using this definition, we find $D(C_2H_4)$ = 2(93) - 12.5 ≅ 174 kcal mol^{-1} and $D(C_2F_4)$ = 2(-49) - (-158) = 60 kcal mol^{-1}. Is this reasonable? That the double bond strength in ethylene is not quite twice that of ethane is sensible: π-bonds are generally weaker than σ-bonds, ethylene (a.k.a. cycloethane) is betrayed as strained (19). That the double bond strength in tetrafluoroethylene is low is also not surprising although -- we recall (8) that fluorinated olefins react both readily and exothermically (though we also recall the caveat: "[while] it is commonly accepted that fluorination [always] destabilizes C=C bonds, ... several results are inconsistent with this conclusion") (20). Difluorocarbene is a particularly stable example of "divalent carbon" albeit far less so than either carbon monoxide or isonitriles in that these latter species are isolable under "normal" laboratory conditions, while their dimers C_2O_2 and RNCCNR have evaded synthesis(21, 22).

It is nonetheless disconcerting to find a double bond that is weaker than a single bond between the same elements. Long before accurate thermochemical data were available from either experiment or theory, J.P. Simons (23, no known relation to the author of ref. 12) noted that the dissociation process of C_2H_4 correlates the reactant olefin and product $2(CH_2)$. That is, the σ and π-orbitals of the former may be transformed into the latter by stretching the carbon-carbon bond, $\sigma^2\pi^2 = 2(\sigma^1\pi^1)$. In contrast, Simons noted that the dissociation of C_2F_4 does not so correlate with its two singlet carbenes, i.e., $\sigma^2\pi^2 \neq 2(\sigma^2)$. The "corrected" bond energy of tetrafluoroethylene is derived by adding to $D(C_2F_4)$ twice the so-called "singlet-triplet gap" of CF_2. [We opt to call this process "Simonizing" (a neologism of one of the current authors, J.F.L., recently used in footnote 105 of reference 22) in due homage to J.P. Simons.] We also use it to reflect the fact that this concept has been reinvestigated by numerous theoreticians, often without due attribution, with the use of an impressive range of rigor and empirical relationships. Relevant references include those of Berson et al. (22), Carter and Goddard (24, 25), Gimarc et al. (26. 27), Hoffmann et al. (28, 29), Ruedenberg et al. (30, 31) and Trinquier and Malrieu (32, 33).] That is, we consider the energy needed to cleave C_2X_4 to properly correlating "single carbon pieces." Accordingly, in equation 5, we define this "Simonized" dissociation energy D^* as

$$D^*(C_2F_4) \equiv D(C_2F_4) + 2\Delta E_{ST}(CF_2) \qquad (5)$$

and recognize that, in fact, $D^*(C_2H_4) = D(C_2H_4)$. But what is $\Delta E_{ST}(CF_2)$ as we also ask for numbers? Admittedly, this quantity is known from experiment (34) and from high quality ab-initio calculations (24, 25) to be some 57 kcal mol^{-1}. By the "perfluoro" and "π-fluoro" effects we would have deduced that $IP_\pi(^3CF_2) \approx IP_\pi(^3CH_2) = 10.40$ eV ($\cong$ 240 kcal mol^{-1}). We note that π-ionization of $^3CF_2$ $(\sigma^1\pi^1)$ and $^1CF_2$ (σ^2) yield the same $^2CF_2{}^+$ (σ^1) ion. The experimental ionization potential of singlet CF_2 is 11.42 eV $\cong$ 263 kcal mol^{-1}. We thus "derive" ΔE_{ST} $(CF_2) \approx 263 - 240 = 23$ kcal mol^{-1}, in contradistinction to the 57 kcal mol^{-1} from the literature. That ΔE_{ST} is so underestimated by this analysis corroborates the earlier related analyses (21, 35) that $^3CF_2$ is destabilized as well as $^1CF_2$ is stabilized. It also corroborates the earlier conclusion that fluorination decreases the π-ionization potential of a radical. In any case, we find that $D^*(C_2F_4) \equiv D(C_2F_4) + 2\Delta E_{ST}(CF_2) = 60 + 2(57) = 174$ kcal mol^{-1}. The bonding and energetics of C_2F_4 and C_2H_4 are perhaps fortuitously identical to within a kcal mol^{-1}: as unstrained, symmetrical olefins we would have expected the bond energies to be merely "similar."

Bond Strengths of Olefin Radical Cations

We have just derived $D^*(C_2F_4) \approx D^*(C_2H_4) \approx 174$ kcal mol^{-1}. As befits the "perfluoro" and "π-fluoro" effects on "normal", closed-shell species, we expect that the ionization potential of tetrafluoroethylene, the parent hydrocarbons, and the various fluorinated species to be "similar." Table II, with all its numbers in eV, documents this.

From the values given earlier, $D(C_2F_4{}^+)$ may be calculated to be 90 kcal mol^{-1} while $D(C_2H_4{}^+)$ likewise is found to be 169 kcal mol^{-1}. Why is $D(C_2H_4{}^+)$ so large and/or $D(C_2F_4{}^+)$ so small? Depending on the dihedral angle between the two CH_2 planes, any nonplanarity of $C_2H_4^+$ results in stabilization albeit by differing degrees of C-C π overlap and C-H bond and hyperconjugation. However, regardless of the geometry of $C_2F_4^+$, we do not expect it to be so stabilized because C-F bonds are not expected to hyperconjugate. Furthermore, consider either of the two equivalent resonance structures for $C_2X_4^+$ (cf. equation 6)

$$\cdot CX_2\text{-}CX_2^+ \longleftrightarrow {}^+CX_2\text{-}CX_2\cdot \qquad (6)$$

For X = H, both the radical and cationic "halves" of the molecule are naturally planar. For X = F, by analogy to the $C_2F_6^+$ story and recalling the geometry of $CHF_2\cdot$ from ref. 10, only the cationic half is. To maximize resonance stabilization (again recall the case of $C_2F_6^+$) both halves of the molecule should have the same geometry. However, whether the radical planarizes and/or the cation pyramidalizes, $C_2F_4^+$ is destabilized. Nonetheless, since the ionization potential of C_2F_4 is experimentally found to be lower than that of C_2H_4, we are hard pressed to invoke destabilization of the radical cation of former: the bond energy difference still seems excessive.

However, let us now extend Simons' logic to gain understanding of olefin radical cations as well as of olefins. More precisely, planar $\sigma^2\pi^1$ $C_2H_4^+$ correlates with its dissociation products $\sigma^1\pi^1$ $^3CH_2$ + σ^1 $^2CH_2^+$ but planar $\sigma^2\pi^1$ $C_2F_4^+$ fails to correlate with σ^2 $^1CF_2$ + σ^1 $^2CF_2^+$. Adding but one $\Delta E_{ST}(CF_2)$ to the earlier sum of 90 kcal mol^{-1} results in the new radical cation bond strength, $D^*(C_2F_4^+)$, of 147 kcal mol^{-1} to be compared with $D^*(C_2H_4^+)$ which by definition is identical to $D(C_2H_4^+)$. The difference between the "Simonized" bond strengths of $C_2H_4^+$ and $C_2F_4^+$ is 22 kcal mol^{-1}, a result that seems more sensible than the earlier value if for no other reason than because the new result is so much smaller.

Classical Ethyl Radicals and Cations

We recall that ethyl cation has a bridged, nonclassical structure, and that the classical $CH_3CH_2^+$ is calculated by high level quantum chemical calculations (36) to be ca. 6 kcal mol^{-1} less stable. Combining this difference with the experimentally measured heats of formation of ethyl cation and of the neutral ethyl radical, we derive the "classical" ionization potential of $CH_3CH_2\cdot$ to be 8.4 eV ≅ 193 kcal mol^{-1}. Consider now the geminally-fluorinated ethyl radicals: $CH_3CF_2\cdot$, $CF_3CH_2\cdot$, and $CF_3CF_2\cdot$. The ionization potentials of the first two radicals, 7.9 eV and 10.6 eV (≅ 182 and 244 kcal mol^{-1}) are consistent with our earlier logic. By analogy to $CH_3\cdot$ and $CF_3\cdot$, were $CH_3CF_2\cdot$ planar it would be expected to have an ionization potential no higher than the parent unfluorinated ethyl radical. However, since radicals become more pyramidal upon replacement of hydrogen by fluorines (10) or by methyl groups (37, 38) the real α,α-difluorinated $CH_3CF_2\cdot$ is without doubt nonplanar (39). We thus deduce that the IP of planar difluorinated radical is lower than ethyl radical.

By contrast, perfluoro/π-fluoro effect reasoning correctly suggests ß-fluorination of ethyl radical will increase the ionization potential because the species is nonplanar. For perfluoroethyl radical, the two fluorination effects on ionization potentials run counter to each other. Experiment does not help us decide which effect dominates because the heat of formation of $C_2F_5^+$ is known no better than ± 15 kcal mol^{-1}. That is, from appearance potentials of this ion from C_2F_5Z with Z = F, I, CF_3 and C_2F_5 we obtain ionization potentials of 228, 218, 208 and 199 kcal mol^{-1} (ca. 9.88, 9.45, 9.02 and 8.63 eV). This is clearly too wide of a spread of values to be of thermochemical use to derive meaningful substituent effects.

Let us turn to the question of C-C bond strengths in ethyl radicals where we will look at the energetics of bond cleavage reaction (7).

$$CX_3CY_2\cdot \longrightarrow CX_3\cdot + CY_2 \qquad (7)$$

Two different quantities, D which is defined as the energy needed to form the ground state radical and carbene and the "Simonized" quantity D^* in which the triplet carbene is utilized so that the two half-filled σ orbitals of the fragments can combine to form the new C-C bond are shown in Table III. While we can offer no explanation for the changes in the bond strengths of the variously substituted radicals upon fluorination, it is nonetheless readily seen that the "Simonized" D^* values have removed most of the wide swings of the values of D associated with increasing fluorination.

Let us turn now to the C-C bond strengths in classical ethyl cations, $CX_3CY_2^+$. We now define two different quantities that are associated with bond strengths. The first is D_{het}, the energy to dissociate the cation "heterolytically" via reaction (8)

$$CX_3CY_2^+ \longrightarrow CX_3^+ + {}^1CY_2 \qquad (8)$$

The second quantity is $D_{hom}(CX_3CY_2^+)$, the energy to break the C-C bond "homolytically" in as reaction (9)

$$CX_3CY_2^+ \longrightarrow CX_3\cdot + CY_2^+ \qquad (9)$$

(We use these terms to convey that the two electrons in the C-C bond are distributed unevenly or evenly respectively. We additionally recognize D_{het} as an "anti-Simonized" process in that we must use the singlet carbene -- and so, we must use the excited σ^2 CH_2 with $\Delta E_{ST} \cong -10$ kcal mol^{-1} (40, 41) -- in order to correlate reactants and products correctly.) Table IV presents the derived numbers, with the warning that the value for the pentafluorethyl cation is suspect. Discernible trends may be seen: the heterolytic bond energy decreases nearly linearly with the number of fluorines while the homolytic bond energy seems to depend on the degree of fluorination on the ß-carbon, but not upon the α-carbon. We do not know enough to ascertain whether these regularities should give us confidence in the value used for the heat of formation of pentafluoroethyl cation.

The Energetics of Vinyl Radicals and Vinyl Cations

Our analysis for vinyl radicals proceeds analogously to the earlier discussion of substituted ethylenes. We recognize that corresponding to the dissociation of ethylenes, there are two processes corresponding to the cleavage of $CX_2CY\cdot$ into CX_2 and CY. The first designated by bond energy D considers formation of ground state CX_2 and CY. The second explicitly assumes both fragments have been "Simonized," i.e. we consider reaction (10) and the associated bond energy D^*

$$CX_2CY \longrightarrow {}^3CX_2 + {}^4\Sigma CY \qquad (10)$$

The only relevant vinyl radicals for which we know the heat of formation are the parent and perfluorinated cases for which the necessary values of ΔE_{DQ}, the doublet-quartet gap, are 17 and 61 kcal mol^{-1} respectively for CH (42) and CF (43). The dissociation energies are presented in Table V. It is seen that the bond strengths parallel those of ethylene and its tetrafluoro analog. Without "Simonizing," the bond strength of the hydrocarbon greatly exceeds that of the fluorinated species. However, making the appropriate "corrections," for spin multiplicity and correlating the 2- and 1-carbon species, the two carbon-carbon bond strengths are nearly the same.

Paralleling the ethyl cation study (36) the parent vinyl cation has a bridged, nonclassical structure, but the classical CH_2CH^+ is calculated by high level quantum chemical calculations to be only 3 kcal mol^{-1} less stable than the observed form. While neither comparable studies nor experimental measurements have been reported on isomeric forms of any of the fluorinated cations, comparison of hydrogen and fluorine-bridging suggests that it is highly unlikely (4) these are nonclassical as well.

The simplest are the heterolytic and homolytic bond cleavage reactions, equations (11) and (12)

$$CX_2CY^+ \longrightarrow {}^2CX_2^+ + {}^2\pi CY \qquad (11)$$

$$CX_2CY^+ \longrightarrow {}^1CX_2 + {}^1\Sigma CY^+ \qquad (12)$$

Table II. Ionization Potentials (in eV) of Ethylene and its Fluorinated Derivatives

n_F	(posns.)	IP	(posns.)	IP	(posns.)	IP
0		10.51				
1		10.36				
2	(1,1)	10.29	(Z:1,2)	10.23	(E:1,2)	10.21
3		10.14				
4		10.12				

Table III. The Bond Strengths (in kcal mol^{-1}) D and D^* for Ethyl Radicals

n_F	(positions)	D	D^*
0		99	99
2	(1,1)	58	115
3	(2,2,2)	107	107
5	(1,1,2,2,2)	54	111

Table IV. The Bond Strengths D_{het} and D_{hom} (in kcal mol^{-1}) for Classical Ethyl Cations

n_F	(positions)	D_{het}	D_{hom}
0		142	143
2	(1,1)	103	140
3	(2,2,2)	78	101
5	(1,1,2,2,2)	46	104

We refer to these cleavages as heterolytic and homolytic because, upon dissociation of the vinyl cation, of the 4 electrons that bound the CX_2 and CY fragments, 1 and 3, and 2 and 2 electrons go respectively with these fragments. The bond energies for these two processes are thus referred to as D_{het} and D_{hom} respectively. Note that neither dissociation process correlates the reactants and products: $\sigma^2\pi^2$ correlates with neither $\sigma^1 + \sigma^2\pi^1$ nor with $\sigma^2 + \sigma^2$. Consider now the "Simonized" products, the triplet state of CX_2 and the triplet ($^3\pi$ $\sigma^1\pi^1$) CY^+ species and the associated bond strength D^*. (The necessary singlet-triplet split is taken from high level ab-initio calculations. See the last section of the chapter for computational details.) Table VI gives these bond strengths for the three vinyl cations for which we have experimental thermochemical data. For both "non-Simonized" processes, the bond strength decreases precipitously with increasing fluorination. Indeed, the bond strength for $C_2F_3^+$ is less than for most carbon-carbon single bonds. Encouragingly, the "Simonized" bond strengths are much more comparable to each other. Additionally, it is not surprising that ß-fluorination destabilizes vinyl cations so that D^* is smaller for the perfluorovinyl cation than the other two ions.

The Energetics of Acetylenes and Acetylene Radical Cations

The bond energies of acetylene and its mono and difluorinated derivatives can be studied in a related manner. Table VII presents the results where D and D^* are the "uncorrected" and "Simonized" dissociation energies corresponding to formation of the doublet $\sigma^2\pi^1$ $^2\pi$ and quartet $\sigma^1\pi^2$ $^4\Sigma$ methylidynes as shown in reactions (13) and (14).

$$XC{\equiv}CY \longrightarrow {}^2\pi\ CX + {}^2\pi\ CY \quad (13)$$

$$XC{\equiv}CY \longrightarrow {}^4\Sigma\ CX + {}^4\Sigma\ CY \quad (14)$$

"Simonizing" results in the bond strengths for the three acetylenes being nearly the same. Admittedly, we find the value of D^* for these acetylenes astonishingly high. We suspect our incredulity is due in large part to the fact that these values are higher than for the likewise triply bonded diatomic nitrogen for which the uncorrected and "Simonized" bond strengths are the same, namely 226 kcal mol^{-1}. Our prejudice perhaps is strengthened because we view N_2 as essentially inert and know that acetylene can burn and explode. However, we know that the reactivity and kinetic instability need not parallel thermochemical stability, and that acetylenes rarely react by C-C bond cleavage (44).

Table V. The Bond Strengths D and D^* (in kcal mol^{-1}) for Vinyl Radicals

n_F	(positions)	D	D^*
0		172	189
3	(1,2,2)	58	197

Table VI. The Bond Strengths D_{het}, D_{hom}, and D^* (in kcal mol^{-1}) for Classical Vinyl Cations

n_F	(positions)	D_{het}	D_{hom}	D^*
0		204	221	236
1	(1)	165	146	247
2	(1,2,2)	86	33	200

We conclude our study with a comparison of the $^2\pi$ $\sigma^2\pi^3$ radical cations of the three acetylenes just discussed. The following dissociation processes will be considered here: the formation of ground state products, $^1\Sigma$ CX^+ + $^2\pi$ CY, and the two sets of products from "Simonized" reactions, $^3\Sigma$ CX^+ + $^2\pi$ CY and $^3\pi$ CX^+ + $^4\Sigma$ CY. These have associated bond strength D, D^* and D^{**}, listed in this order because they correspond to no, one and both products being in their excited state. However, it should be noted that $^3\Sigma$ CX^+ corresponds to two electrons being excited from the ground state so that the degree of excitation is really rather comparable for the two cases. Recall π-ionization is energetically less "expensive" for CF than for CH (i.e., the ionization potential of CF is less than CH). As expected from perfluoro/π-fluoro effect reasoning, fluorination raises the energy of σ-excitation -- recall the $^2\pi$-$^4\Sigma$ split of CH and of CF. In the comparison of monofluoroacetylene with its all-hydrogen and all-fluorine derivatives, it is desirable to consider separately the charged fragment CX^+ as corresponding to two cases X = H and X = F. These are denoted as n_F = 1H and 1F respectively in Table VIII.

The dissociation process forming $^3\pi$ CX^+ + $^4\Sigma$ CY and its associated bond strength D^{**} is the most uniform for the acetylene radical cation cleavage reactions. Encouragingly, this corresponds most closely with the initial "Simonized" dissociation of ethylenes -- recall that neither we, nor Simons, considered the formation of two carbenes, one apiece in their π^2 and σ^2 states upon the dissociation of any ethylene.

Conclusion

It is thus seen that the π-fluoro effect, the assumption that ionization potentials of planar molecules are unaffected by the degree of fluorination, provides a simple conceptual method of ameliorating and understanding the substituent effects of fluorine. Relatedly, the process of "Simonizing" bond strengths, the process that ensures that the σ and π orbitals of a molecule and its cleavage products can be transformed into each other by stretching the carbon-carbon bond, provides a useful procedure. It thus appears that the energetics of a variety of reactive intermediates and their normal counterparts can be understood as a largely coherent whole.

Theoretical Details

Our ab initio quantum chemical calculations were carried out on a VAX 8650 computer with use of the Gaussian 88 system of programs(45). Bond lengths were optimized to 0.001 Å at the HF/6-31G$^{*(*)}$ level of theory with the force relaxation method (46, 47).

Table VII. The Bond Strengths D and D^* (in kcal mol^{-1}) for Acetylenes

n_F	D	D^*
0	231	265
1	177	258
2	117	239

Table VIII. The Bond Strengths D, D^* and D^{**} (in kcal mol^{-1}) for Acetylene Radical Cations

n_F	D	D^*	D^{**}
0	213	320	255
1F	164	271	259
1H	128	290	262
2	69	231	240

Electron correlation was taken into account via fourth order (MP4SDTQ) Møller-Plesset theory (48, 49) at the above bond lengths within the frozen-core approximation (49). For the quartets the spin contamination is small, while for the triplets the spin excitation value, <S^2>, is as high as 3.0 (theoretical, 2.0). However, the single spin annihilation method of Schlegel (50) removed spin contamination in all cases. Table IX presents for all of the diatomic species of interest the bond lengths (in Å) and associated total energies, E_{tot} (in au, where 1 au = 2IP(H) ≅ 627.7 kcal mol^{-1}).

Table IX. Quantum Chemical Calculated Bond Lengths (in Å) and Total Energies (in au) for CH, CF and Their Corresponding Cations

Diatomic	State	r_{CX}	E_{tot}
CH	$^2\pi$	1.108[a]	-38.36695
CH	$^4\Sigma$	1.071	-38.35314
CH^+	$^1\Sigma$	1.105[a]	-37.99412
CH^+	$^3\pi$	1.102	-37.95622
CH^+	$^3\Sigma$	1.241	-37.82315
CF	$^2\pi$	1.267[a]	-137.44867
CF	$^4\Sigma$	1.317	-137.32361
CF^+	$^1\Sigma$	1.144[a]	-137.13257
CF^+	$^3\pi$	1.201	-136.95668
CF^+	$^3\Sigma$	2.081	-136.68696

a See reference (51).

Literature Cited

(1) L.C. Allen, J. Amer. Chem. Soc., **111**, 9003 (1989).
(2) G. Frenking and D. Cremer, Structure & Bonding, **73**, 17 (1990).
(3) S.G. Lias, J.E. Bartmess, J.F. Liebman, J.L. Holmes, R.D. Levin and W.G. Mallard, "Gas-Phase Ion and Neutral Thermochemistry," J. Phys. Chem. Ref. Data, **17**, supplement 1 (1988).
(4) T.D. Thomas, D.C. McLaren. D. Ji and T.H. Morton, J. Amer. Chem. Soc., **112**, 1427 (1990).
(5) C.R. Brundle, M.B. Robin, N.A. Kuebler and H.A. Basch, J. Amer. Chem. Soc., **94**, 1451 (1972).
(6) C.R. Brundle, M.B. Robin and N.A. Kuebler, J. Amer. Chem. Soc., **94**, 1466 (1972).
(7) J.F. Liebman, P. Politzer and D.C. Rosen, in "Applications of Electrostatic Potentials in Chemistry" (ed. P. Politzer and D.G. Truhlar, Plenum, New York, 1981), pp. 295 - 308.
(8) B.E. Smart, in J.F. Liebman and A. Greenberg, "Molcular Structure and Energetics: Studies of Organic Molecules" (Vol. 3) (ed. J.F. Liebman and A. Greenberg, VCH Publishers, Deerfield Beach, 1986), pp. 141 - 192.
(9) B.M. Gimarc, "Molecular Structure and Bonding: the qualitative molecular orbital approach" (Academic Press, New York, 1979), especially pp. 41 - 50.
(10) R.W. Fessenden and R.H. Schuler, J. Chem. Phys., **43**, 2704 (1965).
(11) E.D. Jemmis, V. Buss, P. Schleyer and L.C. Allen, J. Amer. Chem. Soc., **98**, 6483 (1975).
(12) J.F. Liebman and J. Simons, in "Molcular Structure and Energetics: Chemical Bonding Models" (Vol. 1) (ed. J.F. Liebman and A. Greenberg, VCH Publishers, Deerfield Beach, 1986) pp. 51 - 99.
(13) Y.J. Kime, D.C. Driscol and P.A. Dowben, J. Chem. Soc. Faraday 2, **83**, 403 (1987).
(14) M. Tichy, G. Javahery, N.D. Twiddy and E.E. Ferguson, Intl. J. Mass Spectry. Ion Proc., **79**, 231 (1987).
(15) P. Ausloos, S.G. Lias, and M. Meot-Ner (Mautner), personal communication to the authors.
(16) H. Horn and R. Ahlrichs, J. Amer. Chem. Soc., **112**, 2121 (1990).
(17) J.F. Liebman, Struct. Chem., **1**, 395 (1990).
(18) M.C. Inghram, G.R. Hanson and R. Stockbauer, Intl. J. Mass Spectry. Ion Proc., **33**, 253 (1982).
(19) For discussions of the various "uses" of defining ethylene as cycloethane, see A. Greenberg and J.F. Liebman, "Strained Organic Molecules" (Academic Press, New York, 1978), in particular, pp. 10, 29, 43-44, 59 and 66.
(20) Smart, op. cit., p. 164.

(21) Liebman and Simons, op. cit., 80-83, 88-94.
(22) J.A. Berson, D.M. Birney, W.P. Dailey III and J.F. Liebman, in "Modern Models of Bonding and Delocalization" (ed. J.F. Liebman and A. Greenberg, VCH Publishers, Inc., New York, 1989).
(23) J.P. Simons, Nature, **205**, 1308 (1965).
(24) E.A. Carter and W.A. Goddard III, J. Phys. Chem., **94**, 998 (1986).
(25) E.A. Carter and W.A. Goddard III, J. Amer. Chem. Soc., **110**, 4077 (1988).
(26) J.R. Durig, B.M. Gimarc and J.D. Odom, in "Vibrational Spectra and Structure" Vol. 2 (ed. J.R, Durig, Marcel Dekker, New York, 1975), 1 - 133, especially pp. 98 - 123.
(27) Gimarc, op. cit. (ref. 9), especially 139-142.
(28) R. Hoffmann and R.A. Olofson, J. Amer. Chem. Soc., **88**, 943 (1966).
(29) R. Hoffmann, R. Gleiter and F.B. Mallory, J. Amer. Chem. Soc., **92**, 1466 (1970).
(30) L.M. Cheung, K.R. Sundberg and K. Ruedenberg, J. Amer. Chem. Soc., **100**, 8024 (1978).
(31) L.M. Cheung, K.R. Sundberg and K. Ruedenberg, Intl. J. Quantum Chem., **16**, 1103 (1978).
(32) G. Trinquier and J.-P. Malrieu, J. Amer. Chem. Soc., **109**, 5303 (1987).
(33) J.-P. Malrieu and G. Trinquier, J. Amer. Chem. Soc., **111**, 5916 (1989).
(34) S. Koda, Chem. Phys. Lett., **55**, 353 (1978).
(35) S.G. Lias, Z. Karpas and J.F. Liebman, J. Amer. Chem. Soc., **107**, 6089 (1985).
(36) K. Raghavachari, R.A. Whiteside, J.A. Pople and P.v.R. Schleyer, J. Amer. Chem. Soc., **103**, 5649 (1981).
(37) J.B. Lisle, L.F. Williams, and D.E. Wood, J. Amer. Chem. Soc., **98**, 227 (1976).
(38) P.J. Krusic and P. Meakin, J. Amer. Chem. Soc., **98**, 228 (1976).
(39) Y. Chen, A. Rauk and E. Tschuikow-Roux, J. chem. Phys., **93**, 1187 (1990).
(40) D.G. Leopold, K.K. Murray and W.C. Lineberger, J. Chem. Phys., **81**, 1048 (1984).
(41) D.G. Leopold, K.K. Murray, A.E. Stevens Miller, W.C. Lineberger, J. Chem. Phys., **83**, 4849 (1985).
(42) P.R. Bunker and T.J. Sears, J. Chem. Phys., **83**, 4866 (1985).
(43) F.J. Grieman, A.T. Droege and P.C. Engelking, J. Chem. Phys., **78**, 2248 (1983).
(44) A.D. Clauss, J.R. Shapley, C.N. Wilker and R. Hoffmann, Organometallics, **3**, 619 (1984).
(45) M.J. Frisch, M. Head-Gordon, H.B. Schlegel, K. Raghavachari, J.S. Binkley, C. Gonzalez, D.J. DeFrees, B.J. Fox, R.A. Whiteside, R. Seeger, C.F. Melius, J. Baker, R. Martin, L.R. Kahn, J.J.P. Stewart, E.M. Fluder, S. Topiol and J.A. Pople, Gaussian, Inc., Pittsburgh, PA, 1988.
(46) P. Pulay, Mol. Phys., **17**, 197 (1969).

(47) H.B. Schlegel, S. Wolfe, and F. Bernardi, J. Chem. Phys., **63**, 3632 (1975).
(48) C. Møller and M.S. Plesset. Phys. Rev., **46**, 618 (1934).
(49) J.A. Pople, J.S. Binkley and R. Seeger, Int. J. Quantum Chem. Symp., **10**, 1 (1976).
(50) H.B. Schlegel, J. Chem. Phys., **84**, 4530 (1986).

RECEIVED October 16, 1990

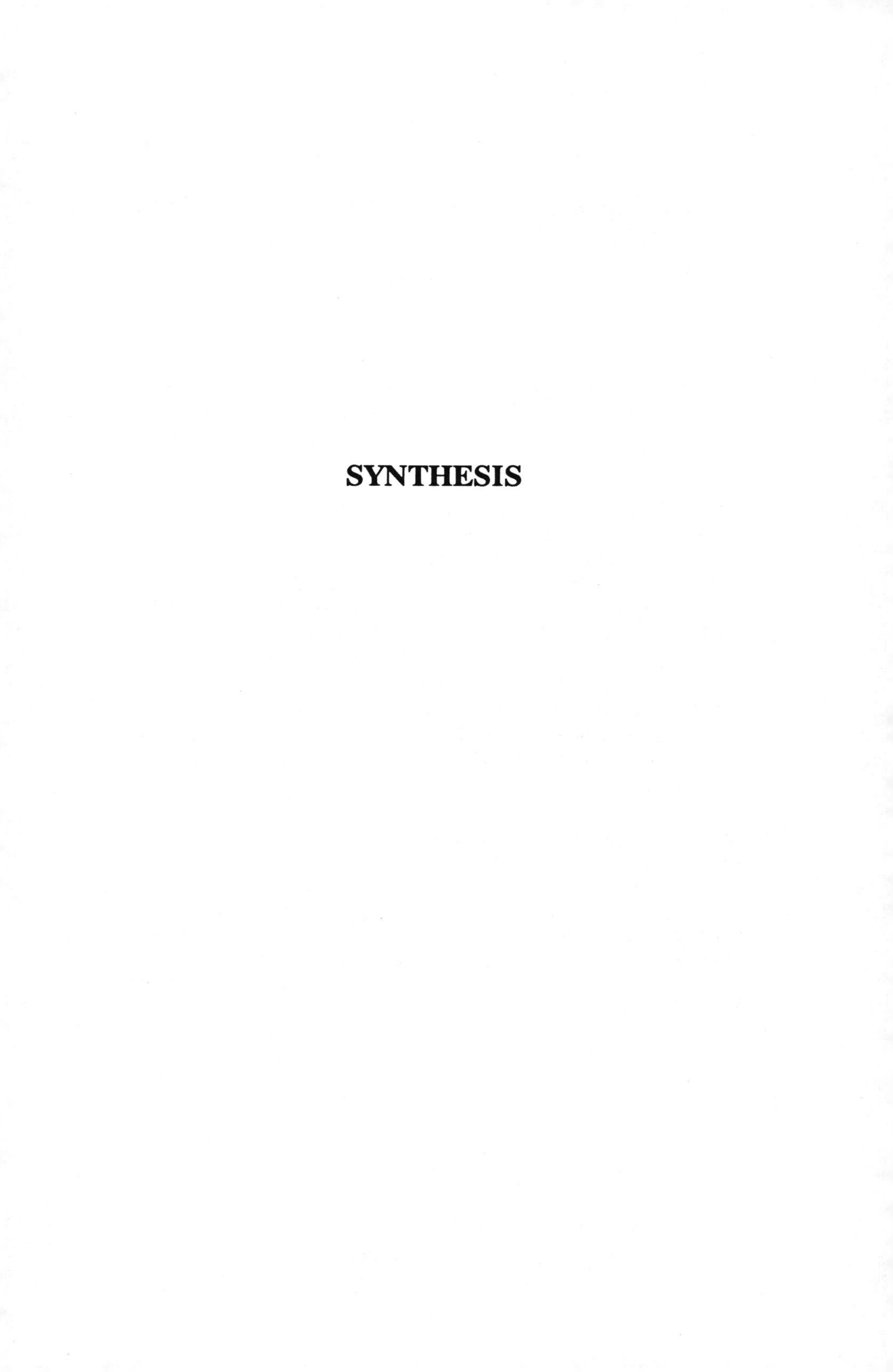

SYNTHESIS

Chapter 4

New Oxidants Containing the O–F Moiety and Some of Their Uses in Organic Chemistry

Shlomo Rozen

School of Chemistry, Raymond and Beverly Sackler Faculty of Exact Sciences, Tel-Aviv University, Tel-Aviv 69978, Israel

The synthesis and chemistry of compounds containing the O-F moiety not attached to perfluorinated alkyl group are described. In general such compounds are less stable than their perfluoroalkyl counterparts. With the exception of AcOF, they should possess at least one electron withdrawing group α to the O-F moiety. It was found that the bigger the alkyl chain of these derivatives the less stable is the compound itself. HOF is also a member of this group and when stabilized by acetonitrile it can perform efficient epoxidation, as well as tertiary hydroxylation on remote and deactivated sites. It also provides an easy way for incorporating various oxygen isotopes such as ^{18}O into organic molecules.

The chemistry of the fluoroxy compounds began with Cady's synthesis of CF3OF more than forty years ago*(1)*. His main idea, of reacting fluorine with fluoro-phosgene in the presence of a dry fluoride serving as a catalyst, was subsequently used by Prager and Thompson and many others who developed several useful variations.*(2- 4)* In the 50's Cady opened another area of fluoroxy chemistry by passing fluorine through a hot mixture of trifluoroacetic acid and water, obtaining trifluoroacetyl hypofluorite - CF3COOF (figure 1).

$$R-\overset{O}{\overset{\|}{C}}-F \xrightarrow{F_2/F^-} RCF_2OF$$

$$CF_3COOH + H_2O + F_2 \longrightarrow CF_3COOF$$

Figure 1. Formation of trifluoroacetyl hypofluorite

0097–6156/91/0456–0056$06.00/0

Apart from the synthesis itself and some studies of physical properties, these compounds did not find much use in organic chemistry for almost 20 years. It was Barton in the late 60's who realized the synthetic potential of CF3OF *(5,6)* and once the direction was set, a flood of works appeared in the literature using this commercial electrophilic fluorinating reagent *(7- 9)*.

Not offered commercially, CF3COOF did not gain as much popularity until the early 80's when we modified its synthesis and showed that it can be an useful reagent in organic chemistry as for example in the preparation of fluorohydrins or α-fluoroketones *(10,11)* -(figure 2).

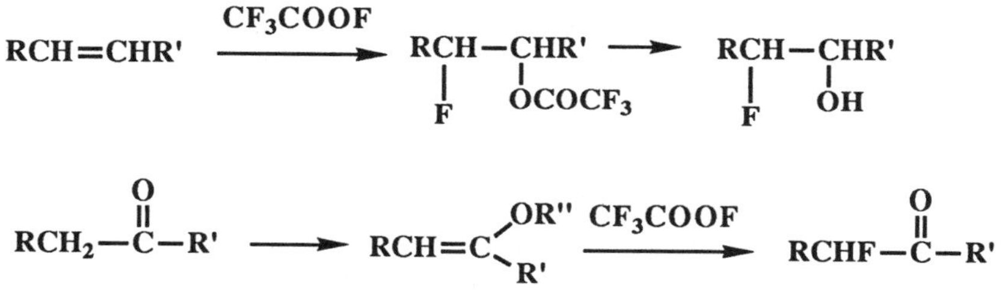

$$RCH{=}CHR' \xrightarrow{CF_3COOF} RCH(F){-}CHR'(OCOCF_3) \longrightarrow RCH(F){-}CHR'(OH)$$

$$RCH_2{-}C(=O){-}R' \longrightarrow RCH{=}C(OR'')(R') \xrightarrow{CF_3COOF} RCHF{-}C(=O){-}R'$$

Figure 2. Use of trifluoroacetyl hypofluorite as a reagent

Later we developed higher homologs in the series of RfCOOF using the same synthetic pathway and used them as unique initiators for polymerization of perfluoro olefins in a way that no undesirable "end groups" were formed *(12)* (figure 3).

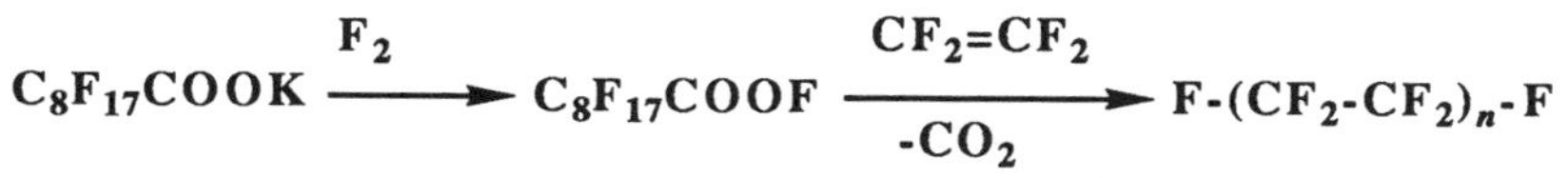

$$C_8F_{17}COOK \xrightarrow{F_2} C_8F_{17}COOF \xrightarrow[-CO_2]{CF_2=CF_2} F\text{-}(CF_2\text{-}CF_2)_n\text{-}F$$

Figure 3. Use of higher homologs as initiators

Acyl Hypofluorites - R_HCOOF

Until recently the carbon skeletons to which the O-F moiety was attached contained no hydrogens and practically always consisted of a perfluoro alkyl group. It was assumed that a hydrogen would trigger an immediate decomposition through an easy HF elimination. A few years ago we showed that there are some exceptions to this "well known fact". We prepared, for the first time, acetyl hypofluorite - AcOF - by reacting sodium acetate with fluorine *(13)* and used it for synthetic purposes without any isolation or purification (figure 4).

$$\begin{array}{l} CH_3COONa \\ \quad or \\ CH_3COOH + NaF \end{array} \xrightarrow{F_2} CH_3COOF$$

Figure 4. Use of acetyl hypofluorite for synthetic purposes

Since its discovery, AcOF has been widely used for fluorination purposes*(14-17)* including positron emitting tomography *(18-19)* and general organic synthesis leading to fluorine free derivatives *(20)*. It should be noted that at the beginning not everyone was fully convinced that the reagent consisted of a single molecule until we *(21)*, but mainly Appelman *(22)*, unequivocally proved its existence.

For several years acetyl hypofluorite remained the only fluoroxy compound containing hydrogens in its alkyl group. Recently we decided to broaden the field by attempting to make additional acyl hypofluorites of this type which might be useful in organic chemistry *(23)*. The first natural choice was to perform a reaction between a cold suspension of sodium propionate and F_2 with or without water, HF and propionic acid. The reason for the addition of the propionic acid and water is that adding acetic acid to AcONa greatly enhances the efficiency of the AcOF production. The corresponding acid and the water have also an essential role in directing the reaction toward hypofluorite formation rather than fluorooxy compounds (see discussion in ref. 10,11). However in each case only traces of oxidizing material were obtained and the only isolable compound was NaF formed quantitatively. It seems that at the reaction temperature of -75 ^{o}C the hypofluorite, if formed, is quite unstable and indeed HF elimination takes place quite easily. It was assumed at the beginning that there are many conformations where either the two α or the three β hydrogens can be very close to the oxygen bound fluorine prompting the elimination. Replacing these hydrogens by methyl groups did not much change the outcome. Sodium pivalate or *t*-butyl acetate produced again only NaF and very volatile components when reacted with fluorine as did also sodium undecanoate.

Trying to avoid too many hydrogens in sodium propionate we replaced the two in the α position with chlorine atoms. This time the reaction with fluorine at -75 ^{o}C resulted in a stable oxidizing solution. The low temperature ^{19}F NMR spectrum revealed a sharp singlet at +134 ppm, a characteristic signal for the COOF fluorine nucleus. Also this oxidizer was reacted with the enol acetate of α-tetralone to give, in 85% yield, the expected α-fluorotetralone, a very characteristic product of electrophilic fluorination (figure 5).

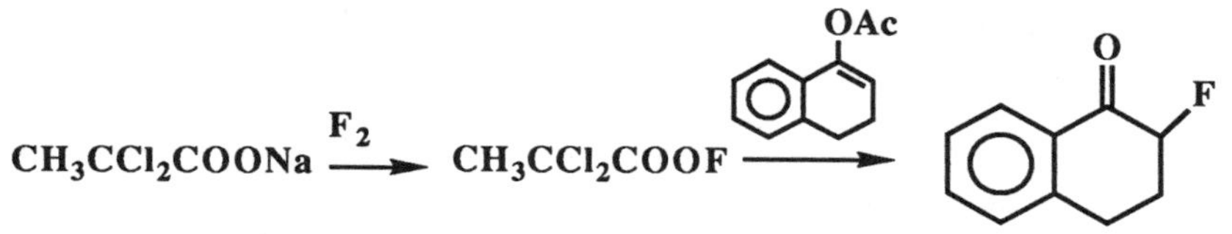

$$CH_3CCl_2COONa \xrightarrow{F_2} CH_3CCl_2COOF \xrightarrow{\text{enol acetate (OAc)}} \alpha\text{-fluorotetralone}$$

Figure 5. Formation of α-fluorotetralone by electrophilic fluorination

The above result indicated that the β hydrogen atoms do not prevent the formation of the O-F bond, but the question remains whether such hypofluorites can be stable in the presence of a hydrogen atom at the α position. The reaction of sodium 2,3-dichloropropionate and sodium 2-chloropropionate with F_2, produced oxidizing solutions, stable at -75 ^oC with physical and chemical properties very similar to those described above. (figure 6). It should be noted here that a single chlorine atom at the β position does not contribute much to the stability of the hypofluorite as is concluded from the case of sodium 3-chloro-propionate which did not produce any oxidizing material when treated with F_2.

$$CH_2XCHClCOONa \xrightarrow{F_2} CH_2XCHClCOOF \quad X = Cl, H$$

$$CH_3CH_2CHXCOONa \xrightarrow{F_2} CH_3CH_2CHXCOOF \quad X = Cl, NO_2, COOEt$$

$$\xrightarrow{\text{enol acetate}} \alpha\text{-Fluoroketone}$$

Figure 6. Formation of α-fluorotetralone from enol acetate

Chlorine however, is not the only substituent at the α position enabling the formation of relatively stable hypofluorites. Sodium 2-nitrobutyrate and monosodium monoethyl 2-ethylmalonate reacted quite efficiently with F_2 to form the corresponding hypofluorites. Both produced the expected α-fluorotetralone from the enol acetate in higher than 85% yield (figure 6).

An important contribution to the stability of the O-F bond is obviously the electronegative group at the 2 position, but it seems that this is not the sole factor which has to be considered. Apparently another important point is the proximity in space to the chain hydrogen atoms of the oxygen bound fluorine. Thus, reacting sodium α-chlorododecanoate with fluorine does not produce any long lasting oxidizing material, although the salt is fully consumed. Similarly with shorter chains of 8 or even 6 carbons, i.e. sodium α-chlorooctanoate and hexanoate, there are numerous conformations where the oxygen bound fluorine can be very close to one of the chain hydrogens (figure 7). It seems that with our reaction conditions, a five carbon chain is the upper limit where hypofluorites can have a synthetically useful life time. Indeed, sodium α- chloropentanoate and monosodium monoethyl propyl malonate react successfully with fluorine to produce the corresponding hypofluorites, although in somewhat lower yield than with the previously mentioned shorter chain homologs (figure 7). What is more, their reaction with enolacetates is also less efficient and the yields of the fluoroketones are only 40 - 50%. This reflects the lower stability of these 5 carbon hypofluorites, where easy decomposition is a competitive route even to the fast reaction with an electron rich enol acetate.

$$CH_3(CH_2)_2CHXCOONa \xrightarrow{F_2} CH_3(CH_2)_2CHXCOOF$$

X = Cl, COOEt *relatively low yields*

$$CH_3(CH_2)_nCHXCOONa \xrightarrow{F_2} NaF + \text{decomposition products.}$$

n = 8; X = H
n = 9; X = Cl
n = 5; X = Cl
n = 3; X = Cl

Figure 7. Formation of hypofluorites

Apart from the fact that such hypofluorites can exist and apart from the academic interest of the inherent limitations of their preparation, there is always the question wether these hypofluorites will have some synthetic value. It is hard for us to predict at this time what the future outcome will be. Neither CF_3OF nor AcOF had found an immediate use in organic chemistry in the first years of their their preparation. We can foresee however, two important areas which may sprung from this subject. The first is broadening the choice of the electrophilic fluorine sources. Thus for example F_2 itself is considered to be the most powerful electrophilic agent, followed by hypofluorites such as CF_3COOF, CF_3OF, CCl_3COOF and AcOF. The differences between these agents however is quite large and new hypofluorites which can be placed between CCl_3COOF and AcOF may increase the selectivity in quite a few cases.

A second and may be more important point is their low thermal stability which may open a new way for decarboxylative fluorination of various acids - a subject which we have started to explore not a long ago (figure 8).

$$RCOONa \xrightarrow{F_2} [RCOOF] \xrightarrow[-CO_2]{\Delta} RF$$

Figure 8. Thermal stability and decarboxylative fluorination of acids

Hypofluorous Acid/Acetonitrile - HOF/CH_3CN

Hypofluorous acid - HOF and its physical properties have been well studied by Appelman during the last 20 years *(24)*. Since however, the HOF stability in solutions is low, its synthetic potential is quite limited *(25)*. We have observed though, that when HOF is created by the action of F_2 on water using acetonitrile as a solvent, a relatively stable oxidizing solution is obtained with a half life time of several hours at room temperature. Some spectroscopic data and chemical reactions indicate that this oxidizing agent is indeed HOF stabilized by complexation with acetonitrile. Thus for example the ^{19}F NMR spectrum of the neat uncomplexed HOF is +20 ppm *(26)* . The same fluorine atom when complexed with acetonitrile resonates at -9 ppm. The complexation process seems to be reversible and it is possible to distill the pure HOF from its acetonitrile complex. When excess of other solvents such as CH_2Cl_2 are being added, the ^{19}F NMR chemical shift shows that the complex becomes weaker up to the point where the ^{19}F NMR signal is practically identical with the one of the neat HOF. This process, as one could expect, is temperature dependent and low temperature slow the process.*(27)*. We have tried quite a few other solvents, but it turns out that CH_3CN is unique in its ability to stabilize HOF, apparently through an effective donor acceptor relationship *(28)*. HOF is the only possible compound were the oxygen atom is always partially positive *(29)*. Stabilizing it enables us to start to explore, synthesis-wise, its potential unique behavior. It should be stated at this point that although the rest of this chapter describes some oxygen transfer reactions, these could never be accomplished without the direct use of F_2 and we would like to stress the point that techniques developed for fluorination purposes can also be employed in general organic chemistry.

Epoxidations with HOF/CH_3CN. The goal of the first set of experiments was exploiting the oxygen's electrophilicity by its reactions with regular nucleophilic olefins. The results were usually fast and high yield epoxidations carried out in a very convenient temperature range of -20 to +20 °C *(30)*. There are, of course, many epoxidation methods to suit the many types of double bonds. Prolonged, high temperature treatment of an olefin with a per-acid is a very popular route. More resistant olefins such as enones, are treated with strongly basic H_2O_2. In many other cases a halohydrin has to be constructed first, followed by a second basic elimination step. Recently a new and very promising method has been developed using dimethyl and methyl trifluoromethyl dioxiranes *(31)*. It seems nevertheless, that having a new "universal" reagent suited to all types of olefins and able to form epoxides at room temperature in a fast and high yield reaction could be highly interesting.

Thus it takes less than a minute for *trans* stilbene to react with the oxidizer at 0 °C producing *trans* stilbene oxide in higher than 90% yield. The physical constants of the oxide show a full retention of the starting material configuration. Similar results were observed with *cis* stilbene, which was converted to *cis* stilbene oxide without any trace of the trans isomer. Substituted stilbenes react as well, as evidenced by the transformation of *trans* 4-chlorostilbene and *trans* 4,4'-dinitrostilbene to the corresponding epoxides in 90% and 70% yield respectively (figure 9).

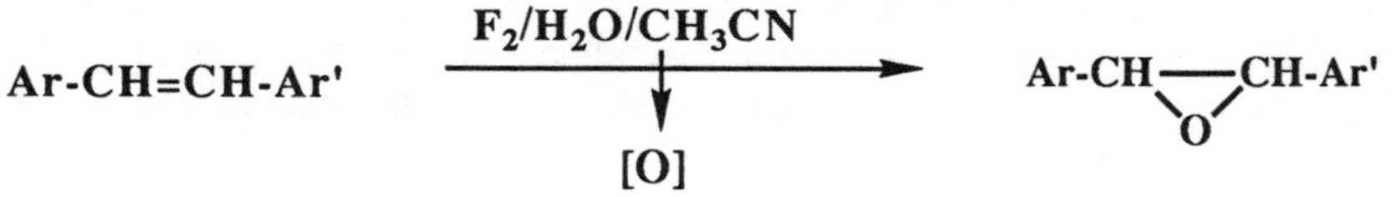

Ar-CH=CH-Ar' —($F_2/H_2O/CH_3CN$; [O])→ **Ar-CH—CH-Ar'** (epoxide, O bridging)

Ar = Ar' = Ph (*trans*)

Ar = Ar' = Ph (*cis*)

Ar = *p*-ClC_6H_4; Ar' = Ph (*trans*)

Ar = Ar' = *p*-$O_2NC_6H_4$ (*trans*)

Figure 9. Transformation of stilbenes to epoxides

This novel epoxidation method is not confined to benzylic olefins. The oxidizing solution, which can be made quickly from commercially available F_2/N_2 mixtures, reacts with cyclooctene at 0 °C in less than a minute to form cyclooctane oxide in 85% yield. Straight chain olefins gave similar results with a full retention of the configuration of the parent alkenes (figure 10).

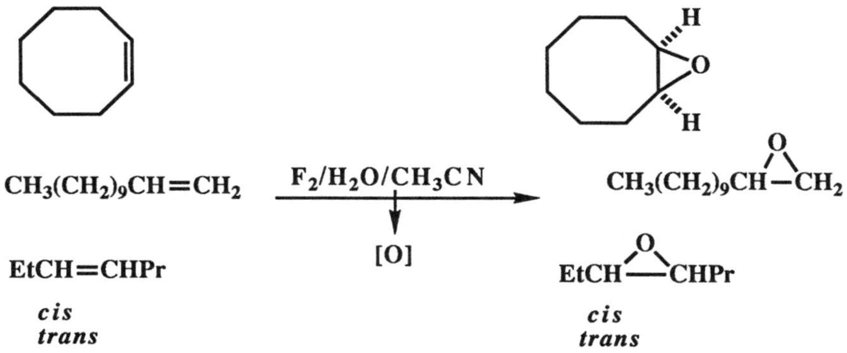

$CH_3(CH_2)_9CH{=}CH_2$ $\xrightarrow{F_2/H_2O/CH_3CN}$ [O] $CH_3(CH_2)_9CH(O)CH_2$

EtCH=CHPr → EtCH(O)CHPr

cis
trans

Figure 10. Epoxidation of straight-chain olefins

Tri-substituted enones are also suitable substrates. 3-Methyl cyclohexenone reacts with a slight excess of the reagent at 0 °C to produce in less then a minute 2,3-epoxy-3-methyl cyclohexanone in 70% yield. The regioselectivity of this electrophilic epoxidation could be clearly demonstrated when the conjugated diene in methyl sorbate was reacted. With only a slight excess of the reagent the γ,δ position was immediately epoxidized producing methyl 4,5-epoxy-2-hexenoate in 90% yield. Using 10-fold excess however results in methyl 2,3:4,5-diepoxy hexanoic acid (as a mixture of two diastereoisomers) obtained in 60% yield (figure 11). It should be mentioned that to the best of our knowledge this type of vicinal diepoxide has never been made before by any direct epoxidation procedure.

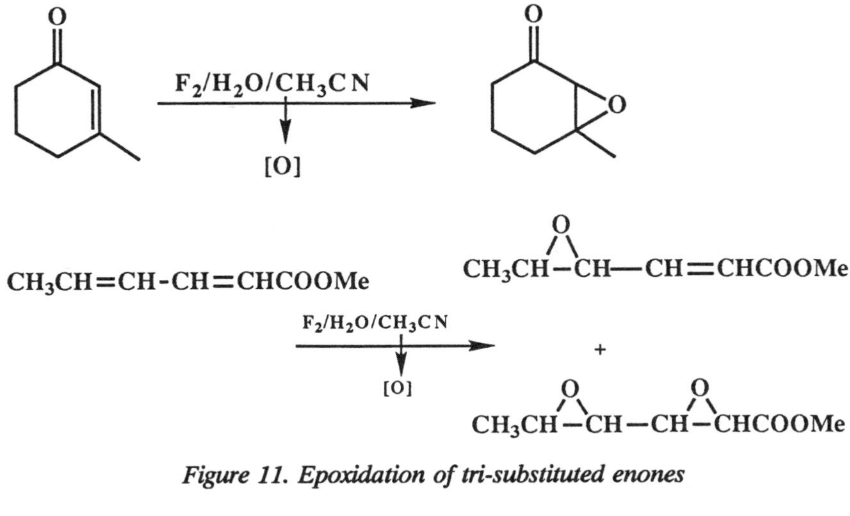

$F_2/H_2O/CH_3CN$ [O]

$CH_3CH{=}CH{-}CH{=}CHCOOMe$ $\xrightarrow{F_2/H_2O/CH_3CN}$ [O] $CH_3CH(O)CH{-}CH{=}CHCOOMe$ + $CH_3CH(O)CH{-}CH(O)CHCOOMe$

Figure 11. Epoxidation of tri-substituted enones

The double bonds in diethyl maleate and fumarate are very electron deficient olefins. These were indeed the most difficult compounds to epoxidize. The usual treatment with slight excess of oxidant for short periods of time did not result in any reaction. With a large excess, however (5-10 molequivalent) progress could be monitored at room temperature and after 12 hours about 90% conversion was achieved. Thus *cis* and *trans* diethyl epoxysuccinate were obtained in 65% and 50% yield respectively. These epoxides are important intermediates in many syntheses, immuno-pharmacological studies and polymerization processes and are always made stepwise by indirect methods. A full retention of configuration was observed and neither epoxide was contaminated with the other (figure 12).

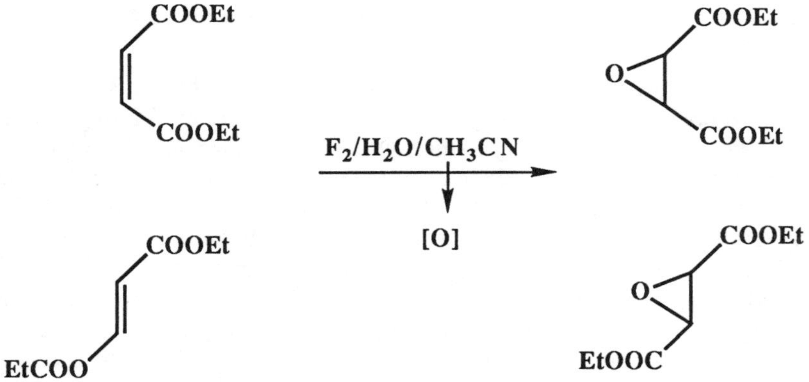

$F_2/H_2O/CH_3CN$ → [O]

Figure 12. Retention of epoxide configuration

This novel epoxidation which uses F_2 as a starting reagent has an additional unique feature. The most convenient source for both ^{17}O and ^{18}O isotopes are the respective water molecules $H_2{}^{17}O$ and $H_2{}^{18}O$. Since the epoxide's oxygen in our method originates directly from the water, one can easily and quantitatively incorporate these isotopes into the desired epoxide and any other products derived from it. Thus we prepared an oxidizing solution using $H_2{}^{18}O$ and reacted it with *cis* stilbene. The oxygen in the resulting cis stilbene oxide indeed contained more then 97% ^{18}O, as could be determined by high resolution mass spectrometry (figure 13).

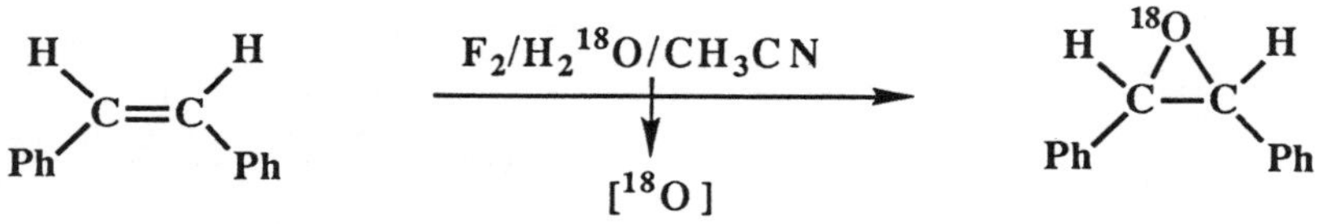

$F_2/H_2{}^{18}O/CH_3CN$ → $[^{18}O]$

Figure 13. Separation of oxygen isotopes by epoxidation

Hydroxylations with HOF/CH_3CN *(32).* Activation of paraffinic and other CH bonds far away from any functional group, is a subject of many recent projects. The more conspicuous approaches are homogeneous catalysis with organometallic complexes and oxygenation processes involving the peroxide bond, including the use of O_3 or H_2O_2 with or without metal cations *(33-36).* Regio- and stereoselective activation is of course a much more demanding process and the two most successful approaches involve Breslow's *(37)* remote radical activation and electrophilic substitution of tertiary hydrogens by F_2 *(38,39).* Since the oxygen atom in HOF/MeCN is strongly electrophilic so, if there are no good "traditional" nucleophilic centers in a molecule, there is a good chance for it to react even with remote deactivated C-H bonds. Of all such bonds the most electron rich are tertiary hydrogens remote from any electron withdrawing moiety *(38).* Indeed when adamantane is treated with the oxidative solution, only a tertiary hydrogen is substituted resulting in an 80% yield of 1-adamantanol. Other paraffins were also similarly hydroxylated including *trans* and *cis* decalins which gave respectively the *trans* and *cis* 9-decalols in yields higher than 80% (figure 14). It should be noted that since the attack is electrophilic in nature the substitution proceeds with a full retention of configuration *(38,40).*

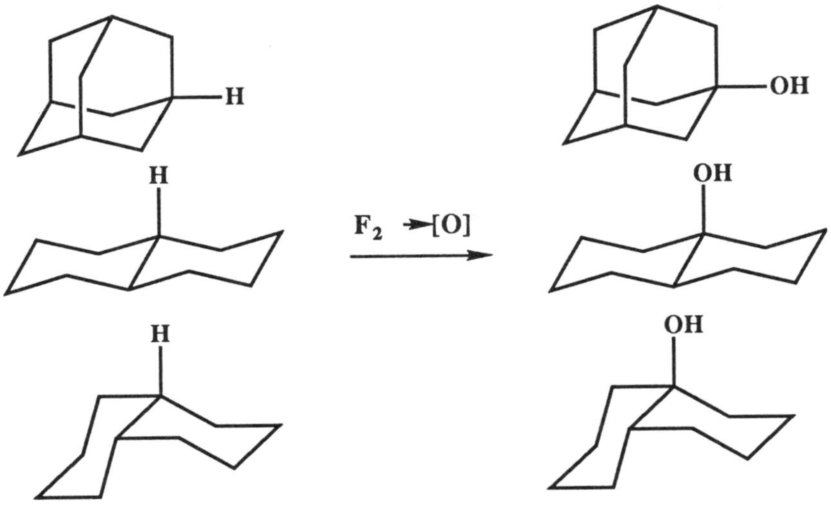

Figure 14. Electrophilic substitution with retention of configuration

As in the case of the epoxidation reaction with HOF/MeCN, this tertiary hydroxylation method can be used for the introduction of the ^{18}O isotope. For example, when adamantane, is treated with the oxidizing solution originated from $F_2/CH_3CN/H_2{}^{18}O$, 1–hydroxy(^{18}O) adamantane was obtained in higher than 80% yield.

Conclusion

In the recent years a new trend in fluoroxy chemistry has been emerging. Alongside traditional perfluorinated reagents some new ones are starting to find their place in the arsenal of organic synthesis. The fact that they lack the powerful electron withdrawing perfluoro-alkyl group make them in some cases gentler fluorinating reagents and in others an appealing tool for preparation of various, sometime difficult to obtain, fluorine free derivatives. In many cases these reagents can be described as fluorine carriers since they are made in situ from F_2 and used without any further isolation and purification.

Acknowledgment: We thank Du Pont Company Wilmington, DE, USA, for supporting this research.

Literature Cited

1. Kellogg, K. B.; Cady, G. H. *J. Am. Chem. Soc.* **1948, *70,*** 3986.
2. Prager, J. H.; Thompson, P. G. *J. Am. Chem. Soc.* **1965, *87,*** 230.
3. Mukhametshin, F. M. *Uspekhi Khimii* **1980, *49,*** 1260.
4. Shreeve, J. M. *Advances in Inorganic Chem. and Radiochem.* **1983, *26,*** 119.
5. Barton, D. H. R.; Godhino, L. S.; Hesse, R. H.; Pechet, M. M. *J. Chem. Soc. Chem. Commun.* **1968,** 804.
6. Barton, D. H. R.; Danks, L. J.; Ganguly, A. K.; Hesse, R. H.; Tarzia, G.; Pechet, M. M. *J. Chem. Soc. Chem. Commun.* **1969,** 227.
7. Hesse, R. H. *Isr. J. Chem.* **1978, *17,*** 60.
8. Purrington, S. T.; Kagen, B. S.; Patrick, T. B. *Chem. Rev.* **1986, *86,*** 997
9. Rozen, S.; Filler, R. *Tetrahedron* **1985, *41,*** 1111.
10. Rozen, S.; Menahem, Y. *J. Fluorine Chem.* **1980, *16,*** 19.
11. Rozen, S.; Lerman, O. *J. Org. Chem.* **1980, *45,*** 672.
12. Barnette, W. E.; Wheland, R. C.; Middleton, W. J.; Rozen, S. *J. Org. Chem* **1985, *50,*** 3698.
13. Rozen, S.; Lerman, O.; Kol, M. *J. Chem. Soc. Chem. Commun.* **1981,** 443.
14. Lerman, O.; Tor, Y.; Hebel, D; Rozen, S. *J. Org. Chem.* **1984, *49,*** 806.
15. Rozen, S.; Lerman, O.; Kol, M.; Hebel, D. *J. Org. Chem.* **1985, *50,*** 4753.

16. Dax, K.; Glanzer, B. I.; Schulz, G.; Vyplel, H. *Carbohyd. Res.* **1987,** ***162,*** 13.
17. Hebel, D.; Kirk, K. L.; Cohen, L. A.; Labroo, V. L. *Tetrahedron. Lett.* **1990,** ***31,*** 619.
18. Diksic, M.; Farrokhzad, S; Colebrook, L. D. *Can. J. Chem.* **1986,** ***64,*** 424;
19. Mislankar, S. G.; Gildersleeve, D. L.; Wieland, D. M.; Massin, C. C.; Mulholland, G. K.; Toorongian, S. A. *J. Med. Chem.* **1988,** ***31,*** 362.
20. Rozen, S.; Hebel, D. *Heterocycles* **1989,** ***28,*** 249.
21. Hebel, D.; Lerman, O.; Rozen, S. *J. Fluorine Chem.* **1985,** ***30,*** 141.
22. Appelman, E. H.; Mendelsohn, M. H.; Kim, H. *J. Am. Chem. Soc.* **1985,** ***107,*** 6515.
23. Rozen, S.; Hebel, D. *J. Org. Chem.* **1990,** ***55,*** 2681.
24. Poll, W.; Pawelke, G.; Mootz, D.; Appelman, E. H. *Angew. Chem. Int. Ed.* **1988,** ***27,*** 392.
25. Andrews, L. A.; Bonnett, R.; Appelman, E. H. *Tetrahedron.***1985,** ***41,*** 781.
26 Hindman, J. C.; Svirmickas, A.; Appelman, E. H. *J. Chem. Phys..* **1972,** ***57,*** 4542.
27 Part of these measurements have been done in collaboration with Dr. E. Appelman from Argonne National Laboratories Il. We hope to publish a full joint report of this work later.
28. Zuchner, K.; Richardson, T. J.; Glemser, O.; Bartlett, N. *Angew. Chem. Int. Ed.* **1980,** ***19,*** 944.
29. Migliorese, K. G.; Appelman, E. H.; Tsangaris, M. N. *J. Org. Chem.* **1979,** ***44,*** 1711.
30. Rozen, S.; Kol, M. *J. Org. Chem.* **1990,** ***55,*** 0000.
31. Murray, R. W. *Chem Rev.* **1989,** ***89,*** 1187.
32. Rozen, S.; Brand, M..; Kol, M. *J. Am. Chem. Soc.* **1989,** ***111,*** 8325
33 Burk, M. J.; Crabtree, R. H. *J. Am. Chem. Soc.* **1987,** ***109,*** 8025;
34. Barton, D. H. R.; Boivin, J.; Lelandais, P. *J. Chem. Soc. Perkin 1* **1989**, 463.
35. Giamalva, D. H.; Church, D. F.; Pryor, W. A. *J. Org. Chem.* **1988,** ***53,*** 3429.
36. Fish, R. H.; Fong, R. H.; Vincent, J. B.; Christau, G. *J. Chem. Soc. Chem Commun.* **1988**, 1504.
37. Breslow, R.; Brandl, M.; Hunger, J.; Adams, A. D. *J. Am. Chem. Soc.* **1987,** ***109,*** 3799.
38. Rozen, S.; Gal, C. *J. Org. Chem.* **1987,** ***52,*** 2769.
39. Rozen, S.; Gal, C. *J. Org. Chem.* **1987,** ***52,*** 4928.
40. Olah, G. A.; Prakash, G. K. S.; Williams, R. E.; Field, L. D.; Wade, K. *Hypercarbon Chemistry;* Wiley: New York, 1987.

RECEIVED October 1, 1990

Chapter 5

Perfluorinated Alkenes and Dienes in a Diverse Chemistry

R. D. Chambers, S. L. Jones, S. J. Mullins, A. Swales, P. Telford, and M. L. H. West

Department of Chemistry, University of Durham, South Road, Durham DH1 3LE, United Kingdom

Model compound studies reveal factors which influence free–radical additions of ethers, polyethers, and crown–polyethers, to fluorinated alkenes. Stereo–electronic effects significantly affect relative reactivities of cyclic amines. Novel dienes are obtained by reactions of some oligomers of fluorinated alkenes with sodium amalgam and some reactions of these systems with nucleophiles are described.

This presentation is deliberately general, i.e. derived from the work of a range of co–workers and their names appear at points in the text corresponding to their work. We are attempting to give a flavour of some of our work, that relates to structure and reactivity, and which illustrates general principles.

Two types of process will be included for discussion because they illustrate a number of fundamentals:

1. Free–radical additions to fluorinated alkenes, (B. Grievson, A.P. Swales, P. Telford, S.L. Jones, A. Joel, R.W. Fuss)
2. Nucleophilic substitutions with fluorinated dienes, (M. Briscoe, M.P. Greenhall, S. Mullins, T. Nakamura)

Reactions with Ethers

Essentially, we are addressing the general problem of use of the carbon–hydrogen bond as a functional group, in the synthesis of fluorine–containing systems and are particularly concerned with free–radical additions of functional compounds to fluorinated alkenes. In principle, therefore, this is an attractive route to various fluorinated derivatives and a considerable amount of earlier work has taken place in this area (see e.g. *1–3*). Indeed, a study of structure and reactivity in free–radical chemistry is of interest in its own right because our understanding of such free–radical processes falls far short of our general knowledge of many ionic processes.

Fluorinated alkenes are particularly reactive towards ethers in free-radical processes; why should this be so? Stabilisation of a radical centre by adjacent oxygen is, of course, well known but if we represent this in

0097–6156/91/0456–0068$06.00/0

Initiation

Chain Transfer

valence–bond terms, then we see why electron–withdrawal may be offset by involvement of the electron pair.

$$X\text{–}\ddot{O}\text{–}CH_2\text{–}R \longrightarrow X\text{–}O\text{–}\dot{C}H\text{–}R \longleftrightarrow X\text{–}\dot{\ddot{O}}^{+}\text{–}\ddot{C}^{-}HR$$

i.e. a *nucleophilic* radical

Consider, for example, the following competition reaction where a deficiency of diethylether reacts *exclusively* with perfluorocyclohexene.

Et_2O $\xrightarrow{\gamma}$

(Deficiency) (B. Grievson)

This reaction demonstrates that fluorinated alkenes are especially reactive towards nucleophilic radicals. Throughout the text, examples of free–radical reactions initiated by γ–rays are interspersed with peroxide–initiated processes and it is worth emphasising that the γ–ray technique is an extremely useful experimental probe because it is, in effect, a universal initiator, i.e. the method can be applied over a wide range of temperature. Furthermore, it leaves essentially no fragments in the product, as do chemical initiators, and the technique may be used with metal apparatus.

A variety of other factors affect reactivity of ethers in additions to fluorinated alkenes and we will gradually explore these. First, a process that was outlined by Muramatsu (*4*), many years ago; a 1,5–intramolecular hydrogen transfer occurs from the intermediate radical [1] and ultimately leads to a di–addition product [2].

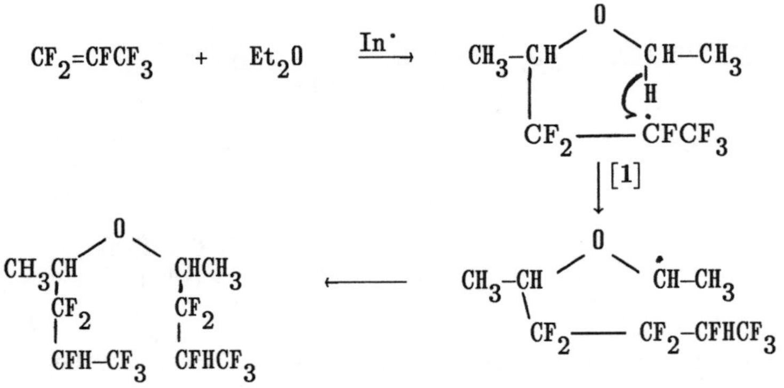

$CF_2{=}CFCF_3$ + Et_2O $\xrightarrow{In^{\cdot}}$ CH_3–CH–O–CH–CH_3 (ring: CF_2—$\dot{C}FCF_3$, H transfer)

↓ [1]

CH_3–CH–O–$\dot{C}$H–CH_3, CF_2— CF_2–$CFHCF_3$

⟵

CH_3CH–O–$CHCH_3$; CF_2 CF_2; CFH–CF_3 $CFHCF_3$

[2] Quant.

Some of the following work was performed with a view to synthesis of perfluoro–ethers and –polyethers, which is not the subject of this present discussion, but we were interested in the relative reactivities of various sites in ether systems. Hence the following results are revealing.

Model Compound Studies (γ–Rays, 20°C) (R_F = CF_2CFHCF_3).

Et_2O : $CF_2{=}CFCF_3$ = 3 : 1

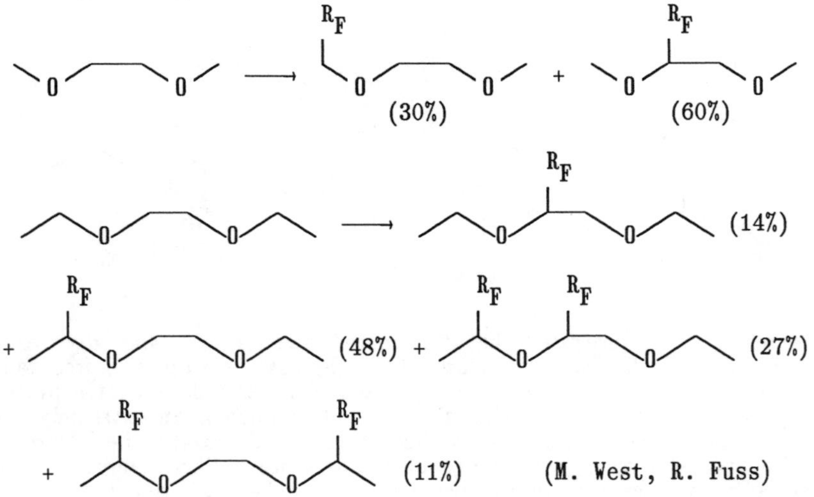

R_F R_F

(30%) (60%)

R_F

(14%)

R_F R_F R_F

(48%) (27%)

R_F R_F

(11%) (M. West, R. Fuss)

Therefore, we may conclude that the relative order of reactivity of various sites in ethers and polyethers, towards fluorinated alkenes in free–radical processes will be:

$$CH_3CH_2O– > –OCH_2CH_2–O– > CH_3O–$$

Increased pressure of the fluorinated alkene and higher temperature, increases the level of substitution and the tri–adduct [3] may be obtained as the major product.

$$\text{EtOCH}_2\text{CH}_2\text{OEt} + CF_2{=}CFCF_3 \xrightarrow[80^o]{(i)} \text{CH}_3\text{CHR}_F\text{OCH}_2\text{CHR}_F\text{OCHR}_F\text{CH}_3 \quad [3]\ (R_F = CF_2CFHCF_3)$$

(i) Dibenzoyl peroxide (DBP), 80°. [**3**] (R_F = CF_2CFHCF_3) (R. Fuss)

We also need to remember that polar effects have a considerable effect on reactivity in free–radical processes and therefore, introduction of a polyfluoroalkyl group should be significant. The question is, what is the effect on reactivity of the resultant ether, of introducing one polyfluoroalkyl group, towards the chain–transfer step, i.e. the ability of an intermediate radical to abstract a hydrogen atom from the ether, to give RH plus an ether derived radical, to continue the chain?

e.g. Additions to $CF_2{=}CFCF_3 \longrightarrow \sim CF_2{-}\dot{C}FCF_3$ ($\equiv R^\cdot$)

$R^\cdot$ + $CH_3CH_2OCH_2CH_3$ vs. $CH_3CHR_FOCH_2CH_3$ $\longrightarrow$ RH etc.

We can obtain information on this process by measuring acetone-butanol ratios in products obtained from the decomposition of di–tertiarybutylperoxide, in the presence of various substrates, using a procedure described previously (*5*). The t–BuO radical may abstract a hydrogen atom, k_2, or lose a methyl radical if the abstraction process is more difficult. Therefore, acetone:butanol ratios are a measure of relative ease of hydrogen abstraction.

$$(t\text{–BuO})_2 \xrightarrow{\Delta} 2(CH_3)_3C{-}\dot{O}$$

$$\xleftarrow{} \dot{C}H_3 + (CH_3)_2CO \xleftarrow{k_1} (CH_3)_3C\dot{O} \xrightarrow[-\overset{|}{\underset{|}{C}}-H]{k_2} (CH_3)_3COH + -\overset{|}{\underset{|}{C}}^\cdot \longrightarrow$$

Substrate (R_F= CF_2CFHCF_3)	Ratio (Acetone : Butanol)
$CH_3CH_2OCH_2CH_2OCH_2CH_3$	0.1
$CH_3CH_2OCH_2CH_2OCHR_FCH_3$	0.9
$CH_3CH_2OCH_2CHR_FOCH_2CH_3$	1.9
$CH_3CH_2OCH_2CHR_FOCHR_FCH_3$	2.5
$CH_3CHR_FOCH_2CH_2OCHR_FCH_3$	75
$CH_3CHR_FOCH_2CHR_FOCHR_FCH_3$	No $(CH_3)_3COH$ detected

(M. West)

Thus, the introduction of polyfluoroalkyl groups has a clear effect in *deactivating* nearby sites to hydrogen abstraction and this can be attributed to polar–effects.

The resulting radical is also subject to polar influences because, being more electrophilic, it will be correspondingly less reactive towards the electrophilic fluorinated alkene.

R_F-substituted ether radical cf. unsubstituted ether radical (more reactive)

($R_F = CF_2CFHCF_3$)

The *converse* of this effect may be achieved when we introduce an electron–donor group and this may be achieved through silyl derivatives. A silyl group attached to oxygen [4] will enhance nucleophilicity of the nearby

$$-\overset{|}{\underset{|}{Si}}\rightarrow\ddot{O}-\overset{\cdot}{\underset{|}{C}}- \longleftrightarrow -\overset{|}{\underset{|}{Si}}\rightarrow\overset{+}{\dot{O}}-\overset{-}{\underset{|}{\ddot{C}}}- \quad [4]$$

$$\text{or } -\overset{|}{\underset{|}{Si}}\rightarrow\dot{C}H_2 \quad [5]$$

site in a radical derived from a siloxane [4], and radicals derived from silanes [5], are also nucleophilic in this context. Some examples of additions to silyl derivatives are shown below.

$$Me_3SiOCH_3 + CF_2{=}CFCF_3 \xrightarrow{\gamma} Me_3SiOCH_2R_F \ (62\%)$$

$$Me_2Si(OCH_2CH_3)_2 + CF_2{=}CFCF_3 \xrightarrow{\gamma} Me_2Si(OEt)—OCH(CH_3)R_F \ (30\%)$$

$$+ \ Me_2Si(OCH(CH_3)R_F)_2 \ (69\%)$$

$$(CH_3)_4Si \xrightarrow{(i)} \underset{[6]}{Me_3SiCH_2CF_2CFHCF_3} + \underset{[7]}{Me_2Si(CF_2CFHCF_3)_2}$$

Ratio [6]:[7] ca. 9:1

(i) DTBP, 140°C (S.L. Jones, A. Swales)

Some regio–isomers [6a,b] are also contained in this mixture and are identified by reaction of the mixture with caesium fluoride, when each is converted to alkene [8a,8b]. This constitutes a novel overall addition of one methylene unit to a fluorinated alkene.

$$Me_3SiCH_2CF_2CFHCF_3 \;[6a] \quad + \quad Me_3SiCH_2\text{-}CF(CF_3)(CF_2H) \;[6b]$$

$$\downarrow CsF \; \Delta \qquad\qquad \downarrow CsF \; \Delta$$

$$CH_2{=}CF\text{-}CFHCF_3 \;[8a] \qquad\qquad CH_2{=}C(CF_3)(CF_2H) \;[8b]$$

(A. Swales, R. Fuss)

Modification of Polyethers

Oxygen atoms in a di– or poly–ether system have two effects: (a) there is a stabilising effect on adjacent radical centres, i.e. activating, as described above but, additionally, the presence of *other* oxygen atoms leads to a net electron-withdrawal and hence deactivation. Therefore, if we separate the oxygen functions we increase the reactivity.

$$-O-(-CH_2CH_2CH_2CH_2O-)_{\overline{n}} \quad > \quad -O-(-CH_2CH_2O-)_{\overline{n}}$$

more reactive

i.e. $-\ddot{O}-\dot{C}H-CH_2 \rightarrow O-$

Stabilising   Reducing reactivity by e–withdrawal

Crown ethers can be modified, e.g. 18–crown–6 gave a range of adducts [10] and, what is remarkable, each of these adducts forms complexes with alkali–metal ions. Indeed, in eluting alkali–metal salts over these adducts, adsorbed on nitrocellulose, and then looking at the systems by plasma–desorption mass-spectrometry, none of the uncomplexed adducts could be detected (Becker, J., Odense University, Denmark, unpublished results). The properties of these unusual complexes will be interesting to explore.

$$\text{18–Crown–6} + CF_2{=}CFCF_3 \xrightarrow[\text{4 Days, Solvent}]{\gamma\text{-Rays}} \text{[18-crown-6]}-(R_F)_n$$

10:1 X.S. C_3F_6

(M. West, A. Joel)   ($R_F = CF_2CFHCF_3$) [10] n = 1–3?

Polyaddition occurs [11] to polyethylene glycol derivatives and note that we deliberately used ethyl derivatives to promote reaction at the end–groups. **Reactions with CF_2=$CFCF_3$ (γ–rays or peroxide initiation).**

$$CH_3CH_2O(CH_2CH_2O)_nCH_2CH_3 \longrightarrow CH_3\underset{R_F}{C}HO(\underset{R_F}{C}H\text{–}CH_2O)_x(CH_2CH_2O)_yO\underset{R_F}{C}HCH_3$$

i.e. n = x + y [11] (R_F = CF_2CFHCF_3)

Typical Results (Conditions depend on scale)

n=	x^a	x^b	Yield (%)	M.Wt.(Calc.)
8–9	4–5	–	86	~1026
~13	9–10	9–10	84	~2146
~45	18	18	80	~4760

(^{a} DTBP, 140°; ^{b} DBP, 80°) X.S. C_3F_6

(P. Telford, M. West, A. Joel)

However, separation of the oxygen functions increases reactivity, as illustrated with polytetrahydrofuran derivatives, where a larger number of polyfluoroalkyl groups are incorporated more easily [12].

$$CH_3CH_2[O(CH_2)_4]_{14}OCH_2CH_3 \qquad (R_F = CF_2CFHCF_3)$$

$$\downarrow \text{X.S. } CF_2\text{=}CFCF_3\text{, DBP, 65}^oC$$

$$CH_3\underset{R_F}{C}H\text{–}[\text{–}O\underset{R_F}{C}H(CH_2)_2\underset{R_F}{C}H\text{–}]_{\bar{x}}[\text{–}O\underset{R_F}{C}H(CH_2)_3\text{–}]_{\bar{y}}O\underset{R_F}{C}HCH_3$$

[12]

Ave. of 20 R_F groups incorporated

↓ Repeat

Ave. of 27 R_F groups incorporated

(M. West)

Workers at ICI, in development of our work, have also modified the system with OH end–groups using this approach (*6*). We can also increase reactivity to some extent by using silyl end–groups; using a silylated polyether [13], where n = ave. ~13, then a product of general formula [14] was obtained, where y – ca. 8. Hydrolysis of the silyl end–groups leads to the corresponding diol. In the absence of silyl, or ether end–groups, the products are complex and include some product of nucleophilic attack. However, solvent can partly overcome this difficulty

$$Me_3Si[OCH_2CH_2]_nOSiMe_3 \xrightarrow[\gamma]{CF_2=CFCF_3} Me_3Si[OCH_2CH_2]_x[OCH(R_F)CH_2]_yOSiMe_3$$

[13] [14] (where $R_F = CF_2CFHCF_3$)

and when γ–rays are used for initiation, in acetone solvent, then polyaddition products of polyethyleneglycols, with hexafluoropropene, may be obtained.

Amines

The same principles that were outlined for ethers, apply to amines but we have a possible competing opportunity for nucleophilic attack (*7*). Nevertheless, the results given below indicate that the free–radical process competes very effectively, i.e. free–radical additions may compete successfully with nucleophilic attack, depending on the system. Other workers have also observed free–radical additions of fluorinated alkenes to amines (*8,9*).

Additions of CF_2=$CFCF_3$ to Amines (γ–Rays)

N-methylpyrrolidine ⟶ 2-R_F-N-methylpyrrolidine ($R_F = CF_2CFHCF_3$)

N-methylpiperidine ⟶ 2-R_F-N-methylpiperidine (81%)

cf. $-\ddot{N}: + CF_2=CFCF_3 \longrightarrow -N^+-CF_2\bar{C}FCF_3$ etc.

(B. Grievson, S.L. Jones)

Stereoelectronic Effects. The conformation of a ring effects the reactivity of cyclic ethers and we have already described the evidence for such processes in the literature (*10*). Here, however, we demonstrate that stereoelectronic effects may influence the reactivity of amines in free radical processes. Using a series of cyclic amines [15], radical addition to a fluorinated alkene may occur but then a choice is possible: (a) the intermediate radical could react with amine to give the 1:1 adduct [16] (k_1) or (b), a 1,5–shift of hydrogen can occur (k_2) which ultimately leads to the 2:1 adduct [17]. We anticipate that k_2 would vary little with the amine but, of course, k_1 will vary with the ease of hydrogen–atom abstraction from the amine. Correspondingly, we find that the ratio of the 2:1 adduct [17] increases substantially in the series

(Additions to CF_2=$CFCF_3$; γ-ray initiators)

N–CH_2CH_3 [15] ⟶ N–CH_2CH_3 (–$CF_2\dot{C}FCF_3$) $\xrightarrow[k_1]{\text{(a) Amine}}$ N–CH_2CH_3 (–CF_2CFHCF_3) [16] 1:1

k_2 ↓ (b) 1,5–Shift

N–$\dot{C}HCH_3$ (–CF_2CFHCF_3) $\xrightarrow{\text{etc.}}$ N–CH_3CH–CF_2CFHCF_3 (–CF_2CFHCF_3) [17] 2:1

N–CH_3 (5-membered ring) > N–CH_2CH_3 (7-membered ring) > N–CH_2CH_3 (6-membered ring)

1: 1–75%	1:1 –46%	1:1 –19%
2: 1–21%	2:1 –52%	2:1 –76%

(S.L. Jones)

shown, therefore demonstrating quite clearly that k_1 varies in this series, with the 5–membered ring system being the most reactive, as was shown for ethers. It may be concluded, therefore, that the order of reactivity reflects the order of *increasing* value of the dihedral angle θ, between the lone–pair on nitrogen and the breaking C–H bond during the hydrogen–abstraction process.

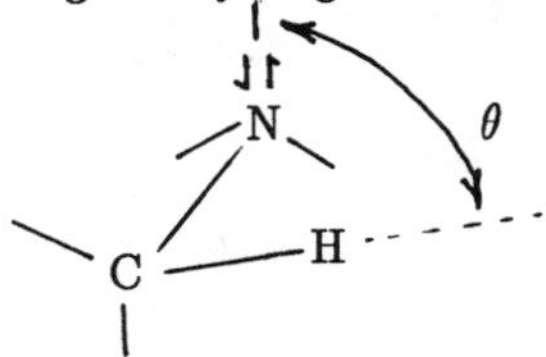

When θ is small, then effective overlap between the lone–pair and the developing radical centre occurs, thus leading to a more stable intermediate radical.

Nucleophilic Reactions of Some Fluorinated Dienes

As is well appreciated, fluorine–containing organic compounds do not occur in nature to any significant degree and, therefore, processes that make carbon–carbon bonds from readily available fluorine–containing compounds, are extremely important. Oligomers of tetrafluoroethene (*11*), perfluorocyclopentene (*12*) and perfluorocyclobutene may be obtained by fluoride–induced processes, although the latter are more efficiently synthesised using a pyridine-induced reaction (*12,13*), and we have now been able to form dienes from some of these systems, using sodium amalgam (*14*). For example, the tetramer **[18]**

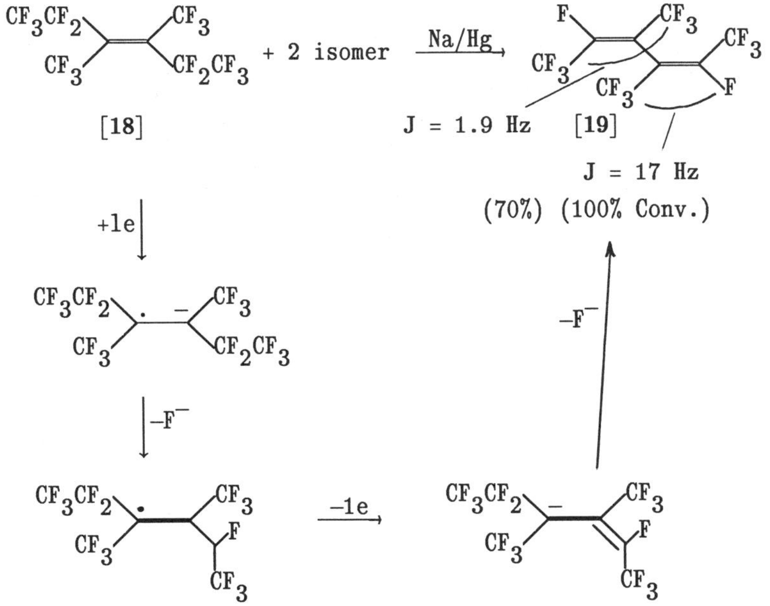

gave the diene **[19]** as the only volatile product, in a reaction that may be reasonably described as involving one–electron–transfer steps, with subsequent elimination of fluoride ion from the intermediate radical anions. As far as we can establish, by F–19 nmr, the diene [19] exists in the *trans–trans* configuration i.e. showing a small *trans* CF_3–CF_3 coupling and larger *cis* CF_3–F coupling.

In an analogous way, both the bicyclopentylidine and bicyclobutylidine derivatives [20] and [21] gave the corresponding dienes [22], [23], using amalgam. Separation of [22] and [23] from their precursors was easily achieved

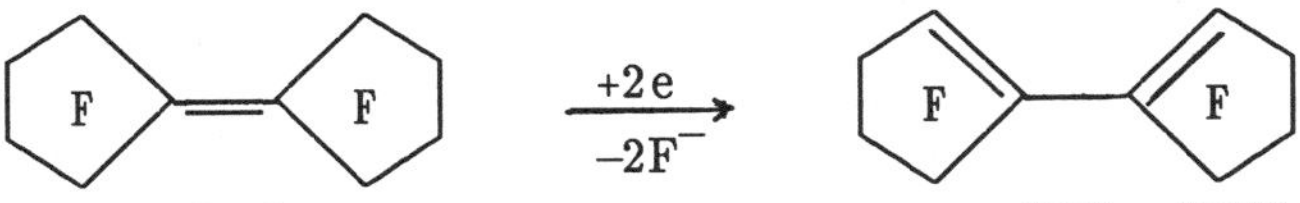

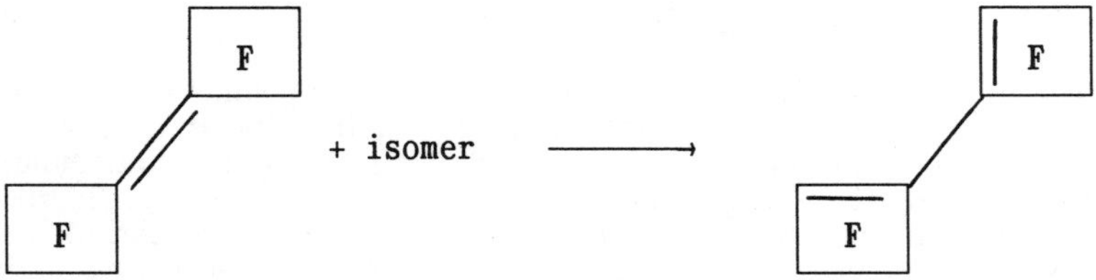

F F + isomer ⟶ F F

[21] [23] (60%, 90% Conv.)

by low–temperature crystallisation of the dienes. Thus, we have a series of novel dienes to explore, and some examples of nucleophilic reactions follow.

Reaction with neutral methanol at room temperature is very revealing. Remarkably, [23] reacted exothermically, while [20] reacted over a period of days, in each case giving the corresponding bis(methylether) by vinylic displacement of fluoride ion. Diene [19], however, required added base to promote the corresponding reaction with methanol. This dramatic difference in

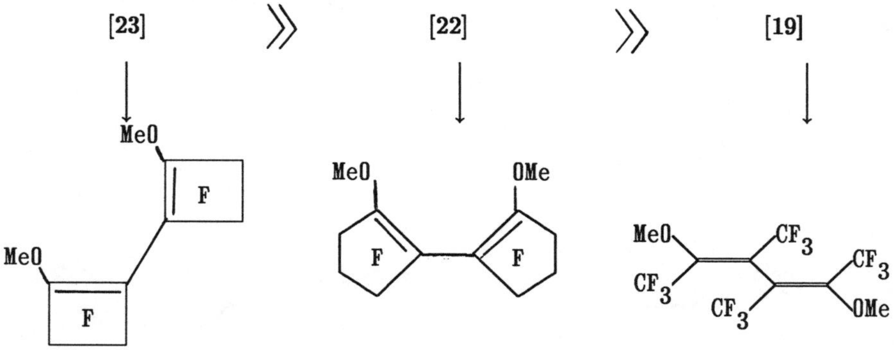

[23] ≫ [22] ≫ [19]

MeO F MeO F | MeO OMe F F | MeO CF_3 CF_3 CF_3 CF_3 OMe

reactivity provides a nice illustration of the effect of angle strain on nucleophilic attack at unsaturated carbon.

Preliminary investigations demonstrate that [19] is an excellent reactant for forming heterocycles. Hydrolysis of [19] occurs quantitatively, to give the known (*15*) furan derivative [24]. Indeed, it is difficult to avoid forming some of the furan unless solvents are kept scrupulously dry. Obviously, for the cyclisation to proceed, then there must be interconversion of isomers and we can think of alternate processes for forming [24], as shown.

An analogous process occurs with K_2S, to form the corresponding thiophene derivative [25], which had previously been obtained from the less accessible (to us) hexafluoro–2–butyne (*16,17*).

Remarkably, we can even form a pyrrole derivative [26], in an efficient process, if we have caesium fluoride present, to avoid loss of fluorine from adjacent trifluoromethyl in the intermediate [27]. In the absence of caesium fluoride, the product [28] is obtained, along with the pyrrole derivative [26] and the formation of [28] is rationalised, as shown.

[19] $\xrightarrow{H_2O/Et_2O}$ (Quant.)

CF3 CF3 CF3 CF3 O

F CF3 CF3 CF3 O⁻ CF3

CF3 CF3 F CF3 ⁻O CF3

CF3 CF3 CF3 CF3 F O

$\longrightarrow$ [24]

[19] $\xrightarrow[DMF,\ RT]{K_2S}$ CF3 CF3 CF3 CF3 S + [24]

[25]

[19] + $PhNH_2$ $\xrightarrow[7d., RT]{CsF}$ [26] (72%)

[27] $\xrightarrow{-HF}$ ... $\xrightarrow{-HF}$ ... $\xrightarrow[(-HF)]{PhNH_2}$ ... $\xrightarrow{-HF}$ [28]

The pyrrole derivative [26] has previously been obtained in an approach that involved a valence isomer of [25] (*18*).

The examples discussed here, illustrate, we hope, some of the diverse factors influencing reactivity of fluorinated–alkenes and –dienes.

Acknowledgments

We thank the United States Air Force Office of Scientific Research, The Science and Engineering Research Council, The European Economic Community, the Electricity Council, and ICI plc, for financial support in various areas of this work.

Literature Cited

1. Starks, C.M. *Free Radical Telomerisation*; Academic Press: New York, N.Y.; 1974, and references cited.
2. Chambers, R.D. *Fluorine in Organic Chemistry*; Wiley–Interscience: New York, N.Y.; 1973, pp 173–178, and references cited.
3. Chambers, R.D.; Grievson, B. *J. Fluorine Chem.* **1985**, *30*, 227, and earlier parts of a series.
4. Muramatsu, H.; Kimoto, H.; Inukai, K. Bull. Chem. Soc. Japn., **1969**, *42*, 1155.
5. Chambers, R.D.; Grievson, B.; Kelly, N.M. *J. Chem. Soc. Perkin Trans. 1*, **1985**, 2209.
6. Powell, R.L.; Young, B.D. *Eur. Pat. Appl.* EP,260,846 **(1988)**; *CA* 1988, *109*, 38488x.
7. Chambers, R.D.; Mobbs, R.H. *Adv. Fluorine Chem.*, **1965**, *4*, 50.
8. Liska, F.; *Coll. Czech. Chem. Commun.*, **1971**, *36*, 1853.
9. Liska, F.; Kubelka, V. *Coll. Czech. Chem. Commun.*, **1971**, *36*, 1381.
10. Chambers, R.D.; Grievson, B. *J. Chem. Soc. Perkin Trans. 1*, **1985**, 2215.
11. Graham, D.P.; *J. Org. Chem.*, **1966**, *31*, 955.
12. Chambers, R.D.; Taylor, G.; Powell, R.L. *J. Chem. Soc. Perkin Trans. 1*, **1980**, 429.
13. Pruett, R.L.; Bahner, C.T.; Smith, H.A. *J. Am. Chem. Soc.*, **1952**, *74*, 1638.
14. Briscoe, M.W.; Chambers, R.D.; Mullins, S.J.; Nakamura, T.; and Drakesmith, F.G.; *J. Chem. Soc. Chem. Commun.* **1990**, paper 1592J.
15. Chambers, R.D.; Lindley, A.A.; Philpot, P.D.; Fielding, H.C.; Hutchinson, J.; Whittaker, G. *J. Chem. Soc. Perkin Trans. 1*, **1979**, 214.
16. Krespan, C.G. *J. Amer. Chem. Soc.*, **1961**, *83*, 3434.
17. Chambers, R.D.; Jones, C.G.P.; Silvester, M.J.; and Speight, D.B. *J. Fluorine Chem.*, **1984**, *25*, 47.
18. Kobayashi, Y.; Ando, A.; Kawada, K.; Ohsawa, A.; Kumadaki, I. *J. Org. Chem.*, **1980**, *45*, 2962.

RECEIVED October 15, 1990

Chapter 6

Perfluorinated Enolate Chemistry

Selective Generation and Unique Reactivities of Ketone *F*-Enolates

Cheng-Ping Qian and Takeshi Nakai

Department of Chemical Technology, Tokyo Institute of Technology, Meguro, Tokyo 152, Japan

A general, highly regio- and stereoselective method is developed to generate the metal *F*-enolates, $CF_3C(OM)=CFR_f$ (R_f=F, CF_3; M=Li, Na, K), from $CF_3CH(OH)CF_2R_f$. The *F*-enolates thus generated are shown to exhibit unique and wide spectra of reactivity toward a variety of reagents. Of particular interest is that the *F*-enolates show a rather unusual electrophilic behavior toward organometallic reagents, while the *F*-enolates still demonstrate the usual enolate reactivities including aldol reactivity. These reactivities of the *F*-enolates are well accommodated by their ab initio molecular orbital calculations. The unique spectrum of reactivity of β-hydro-and β-alkyl-*F*-enolates, $CF_3C(OLi)=CF\text{-}R$ (R=H, *n*-Bu), is also described.

In sharp contrast to the prominent position of metal enolates in synthetic organic chemistry, the chemistry of perfluorinated enolates (*F*-enolates) remains relatively unexplored, mainly because of the lack of practical methods for their generation. Only a few examples of metal *F*-enolates have been recorded in the literature. (*1*, 2). We believe that perfluoro-enolate chemistry can play an equally important role in organofluorine synthesis (Figure 1). In this paper we wish to describe the recent advances in ketone *F*-enolate chemistry which have been made in our laboratory.

Generation of *F*-Enolates

Generation of Parent *F*-Enolate. Recently we have successfully developed a new, practical method for generating the parent ketone *F*-enolate **1** from commercially available hexafluoroisopropyl alcohol (HFIP) (*3*). The newly-developed method is quite simple (eq 1). Thus, the alcohol is treated with two equivalents of buthyllithium in THF around -40 °C for 2 h, or at 20 °C for 20 min to generate *F*-enolate **1** (M=Li) in essentially quantitative yield. Not unexpectedly, the lithium *F*-enolate is quite stable even at room temperature as determined by ^{19}F NMR spectroscopy: ^{19}F NMR (Et_2O, ex. CF_3CO_2H), δ -8.0 (d, d, J=9.4 and 22.6 Hz, CF_3), +30.3 (br., d, J=88.4 Hz, F trans to CF_3), and +41.3 (br., d, J=88.4 Hz, F cis to CF_3).

0097–6156/91/0456–0082$06.00/0

Generation of sodium (**1**, M=Na) and potassium *F*-enolates (**1**, M=K) is also feasible by successive treatment of HFIP with NaH (1 equiv) and *n*-BuLi (1 equiv) and with KH (1 equiv) and *n*-BuLi (1 equiv), respectively, under similar conditions. The Na- and K-enolates show different ^{19}F NMR spectra from that of the Li-enolate; the δ-values for the CF_3 peak (Et_2O, ex. CF_3CO_2H) are -6.4 ppm for M=Na and -7.0 ppm for M=K. It should be noted that the K-enolate is relatively unstable and gradually decomposes at room temperature. Of special interest is the observation that the rate of *F*-enolate formation from the metal alkoxides significantly varies in the order: M=K>Na>Li. We found that the *F*-enolate yields (after benzoylation) of reactions of the alkoxides with *n*-BuLi conducted at- 70 °C for 4 h were 37% for M=Li, 81% for M=Na, and 91% for M=K. This order is somewhat surprising in view of the known order of the ionic character of the M--O bonds, suggesting that the stronger coordination by F to M in the alkoxide (see Figure 2) makes the elimination faster. The role of the coordination by F to M is evidenced by the independent finding that the *F*-enolate formation from the potassium alkoxide is considerably suppressed by addition of 18-crown-6, which reduces the F--K coordination at least partially. For instance, the reaction of the alkoxide (M=K) with *n*-BuLi in the presence of 18-crown-6 (1.0 equiv) at -78 °C for 4 h was found to afford only 19% yield of the *F*-enolate compared with an 83% yield in the absence of the crown ether.

Selective Generation of β-CF_3-substituted *F*-enolates. We have also examined regio- and stereoselectivity in the dehydrofluorinative generation of *F*-enolates (*4*). Thus, the di(*F*-ethyl)carbinol **3a** and *F*-methyl-*F*-ethylcarbinol **3b** were prepared in ca.70% distilled yields by $LiAlH_4$ reduction of the corresponding *F*-ketone, obtained from R_fCO_2Et and CF_3CF_2I according to the literature (*5*). The selectivities observed in the generation of *F*-enolate **4** from alcohol **3** are summarized in Table I (eq 2). For the lithium *F*-enolates (entries 1-5), only the (Z) isomers are formed in ether, whereas in THF the selectively decreases and in THF/HMPA it reverses to favor the (E) isomer. "Internal" enolate **4b** are produced almost exclusively, irrespective of solvent and metal ion employed, except for the reaction with *n*-BuLi in ether, which gives appreciable "terminal" enolate **5** (entry 3). Notably, (Z)-**4b** (M=K) is formed both regio- and stereospecifically in ether (entry 6).

It should be noted here that the "Internal" enolate formation is favored both kinetically and thermodynamically. Under identical reaction conditions [*n*-BuLi (2.1 equiv), THF, -78 °C, 4 h], 100% of *F*-enolate is formed from $(CF_3CF_2)_2CHOH$ but only 37% from $(CF_3)_2CHOH$. The relative stability of the free enolates (gas phase) is (*E*)-$CF_3C(O^-)=CFCF_3$ (0) > (Z)-$CF_3C(O^-)=CFCF_3$ (5.6) > $CF_2=C(O^-)CF_2CF_3$ (26.3 kcal/mol) from the ab initio calculations (*vide infra*). Control experiments, however, rule out any equilibrations under our reaction conditions, and therefore the observed isomer ratios in Table I are the kinetic product distributions.

The increase in (Z)-stereoselectivity with decreasing solvent coordination to Li (HMPA>THF>ether) indicates that the stereoselectivity is controlled by the extent of internal F--Li coordination in the Li alkoxide intermediate. Thus, it appears likely that the HF elimination proceeds in a *trans* fashion exclusively through species **A** in ether, but partially through **B** in THF and more so in THF/HMPA (Figure 3). Further noteworthy is that a similar (Z):(E) ratio (1:3) to that in entry 5 is observed in the *F*-enol ether formation from $CF_3CH(OMEM)CF_2CF_3$ with LDA in THF, where F--Li coordination cannot be operative.

Generation of Terminal *F*-Enolates. We have also developed an entirely different approach for the generation of "terminal" *F*-enolates without regiochemical ambiguity (*6*). For instance (eq 3), treatment of bromodifluoromethyl *F*-ketone **6**

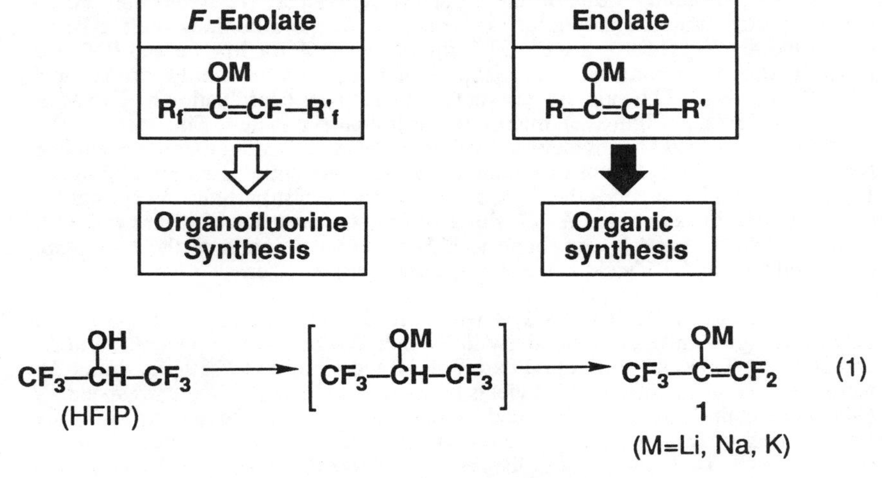

Figure 1. *F*-Enolate *vs*. Enolate.

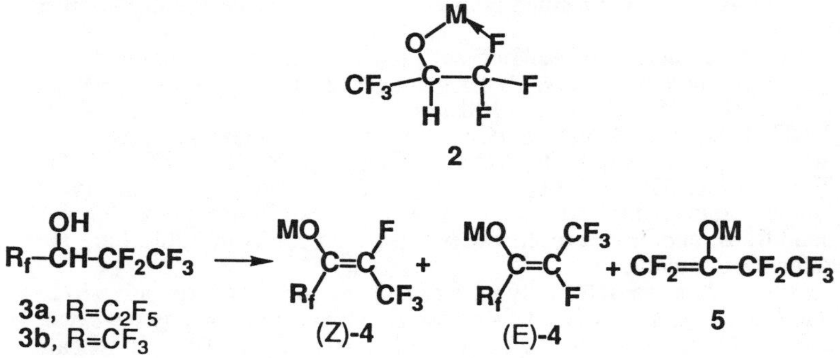

$$CF_3\text{—}CH(O\text{—}M\leftarrow F)\text{—}CF_2F \quad \mathbf{2}$$

$$R_f\text{—}CH(OH)\text{—}CF_2CF_3 \longrightarrow (Z)\text{-}\mathbf{4}:\ (MO)(R_f)C{=}C(F)(CF_3) + (E)\text{-}\mathbf{4}:\ (MO)(R_f)C{=}C(CF_3)(F) + CF_2{=}C(OM)\text{—}CF_2CF_3\ \mathbf{5} \qquad (2)$$

3a, $R=C_2F_5$
3b, $R=CF_3$

Figure 2. Structure of **2**.

Table I. Regio- and Stereoselectivity of *F*-Enolate Formation

entry	alcohol	M[a]	solvent	Isomeric ratio[b] Z-4 : E-4 : 5	yield, %[b]
1	**3a**	Li	Et_2O	100 : 0	81
2		Li	THF	79 : 21	74
3	**3b**	Li	Et_2O	88 : 0 : 12	82
4		Li	THF	87 : 13 : 0	68
5		Li	THF/HMPA[c]	40 : 60 : 0[d]	68
6		K	Et_2O	100 : 0 : 0	78
7		K	THF	65 : 35 : 0	71

[a] Run at -70 °C for 4 h using 2.2 equiv of *n*-BuLi for **4**(M=Li) or 1.2 equiv of KH followed by 1.1 eq. of *n*-BuLi for **4**(M=K). [b] Determined by ^{19}F NMR after trapping with AcCl for **4a** or with MEMCl for **4b**. The isomeric *F*-enol acetates and ethers are clearly distinguishable by ^{19}F NMR (ex. TFA), particularly by the multiplicity of the b-CF_3 signals: d -11.5 (d, t, J=7.5 and 15.1 Hz) for Z-**4a**; d -7.7 (d, J=7.5 Hz) for *E*-**4a**; d -11.6 (br. q, J=7.5 Hz) for Z-**4b**, d -8.6 (d, J=7.5 Hz) for *E*-**4b**; d +7.3 (s) for **5**. [c] THF:HMPA=4:1 by volume. [d] (Z)-**4b**:(E)-**4b**:**5**=38:58:4 from Me_3SiCl trapping experiments.

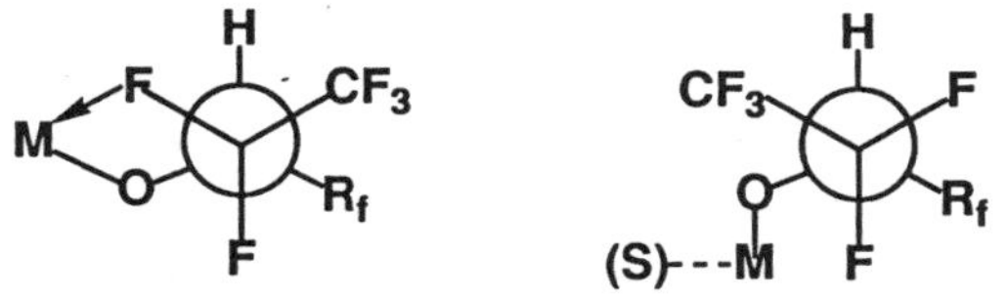

Figure 3. Chelation *vs*. Nonchelation.

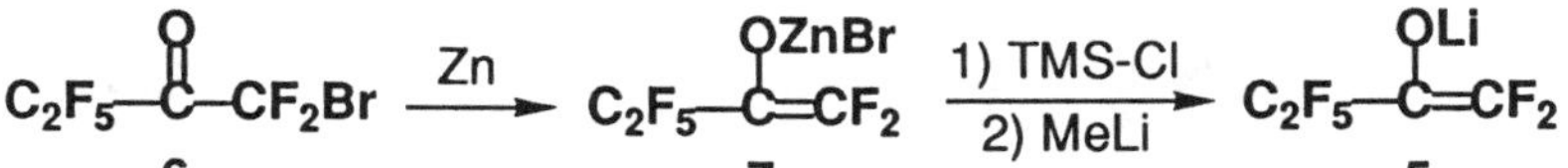

$$\underset{\mathbf{6}}{C_2F_5-C(=O)-CF_2Br} \xrightarrow{Zn} \underset{\mathbf{7}}{C_2F_5-C(OZnBr)=CF_2} \xrightarrow[2)\ MeLi]{1)\ TMS\text{-}Cl} \underset{\mathbf{5}}{C_2F_5-C(OLi)=CF_2} \quad (3)$$

with zinc in THF gives rise to zinc *F*-enolate **7** [^{19}F NMR (THF), δ 27.8 (d,t), and 38.8 (d,t) for the olefinic fluorines], which may be further treated successively with chlorotrimethylsilane and methyllithium to generate lithium *F*-enolate **5** [^{19}F NMR (THF), δ 33.3 (d,t), and 44.0 (d,t) for the olefinic fluorines]. Since various bromodifluoromethyl *F*-ketones are easily prepared from *F*-alkyl iodide and bromodifluoroacetate (both commercially available), this method should be applicable to generation of "terminal" *F*-enolates with different perfluoroalkyl groups.

Reactivities of *F*-Enolates

With the practical methods for the generation of *F*-enolates in hand, we next turned attention to the chemistry of *F*-enolates. The fundamental questions to be answered are as follows. (a) Do *F*-enolates show the ambident nucleophilic reactivity like hydrocarbon enolates ? (b) How about the aldol reactivity ? (c) Do *F*-enolates exhibit any unique reactivities ? In other words, do *F*-enolates show rather the electrophilic reactivity of perfluoroolefins? With these questions in mind, we carried out reactions of *F*-enolates with a wide variety of reagents.

Triple Reactivity of Parent *F*-Enolate. The following scheme (Figure 4) demonstrates the wide spectrum of reactivity observed with the parent lithum *F*-enolate **1** (*3*). Several features of the *F*-enolate reactivity are now revealed: (a) The *F*-enolate exhibits the usual enolate reactivities, *i. e.*, the *O*- and *C*-nucleophilicity, to afford various classes of polyfluorinated compounds (**8-10**) which are otherwise difficult to obtain. (b) In reactions with reagents bearing an active hydrogen, the initial protonation occurs at the β-carbon, not at the oxygen, to yield the products of type **9**. (c) The *F*-enolate is capable of undergoing the aldol reactions with various carbonyl partners to afford the adducts as hydrates **10**. (d) Most significantly, the *F*-enolate exhibits a rather unusual electrophilic reactivity toward organometallic reagents to generate a geometric mixture of the β-alkyl *F*-enolates (**11**) via an addition-elimination process. This means that the *F*-enolate behaves as a perfluoroolefin which is well known to undergo a similar type of substitution reaction. The interesting "triple reactivity" of the *F*-enolate will be discussed on the basis of theoretical calculations in the subsequent section.

Reactivity of β-CF_3-Substituted *F*-Enolates. The following scheme (Figure 5) shows the characteristic spectrum of reactivity observed with the β-CF_3-substituted *F*-enolate **4b** (M=Li) (*4*). (a) The β-CF_3-*F*-enolate reacts with (methoxyethoxy)methyl chloride (MEMCl) to afford the *O*-alkylation product (**13**) as described for the parent *F*-enolate **1**. (b) The reaction with water is also the same as reported for **1** to give the *C*-protonated product (**14**). (c) Interestingly, however, the *F*-enolate **4b** no longer undergoes aldol reaction with any carbonyl partners, not even with highly reactive CF_3CHO. This is in sharp contrast to the relatively high aldol reactivity observed for the parent *F*-enolate. (d) More significantly, the electrophilic reactivity of **4b** toward *n*-BuLi is remarkably higher than that of parent **1** to give, after benzoylation, the butylated *F*-enol ester (**15**) in 55% yield, as compared with an only 28% yield of **12** from **1** under the same conditions. Thus, it is safe to conclude that the β-CF_3-substitution on *F*-enolate decreases the aldol reactivity (*C*-nucleophilicity) but enhances the *C*-electrophilicity toward organometallic reagents.

Theoretical Calculations. Both the unique triple reactivity of the parent *F*-enolate and the significant difference in reactivity between the parent (**1**) and β-CF_3

O-E
CF_3-C=CF_2
8

E-Cl

OH
CF_3-C-CF_2H
Nu
9

NuH

OLi
CF_3-C=CF_2
1

R-M

[OLi
CF_3-C=CF-R
11]

R'COR"

R'COCl

O OH
CF_3-C-CF_2-C-R"
10 R'

OCOR'
CF_3-C=CF-R
12

E-Cl (product yield): PhCOCl (99%), Me_3SiCl (80%), $(MeO)_2SO_2$ (72%)
NuH (product yield): H_2O (86%), $PhCH_2OH$ (71%), $PhCONH_2$ (74%), $CHF(CO_2Et)_2$ (61%)
R'COR" (product yield): CF_3CHO (87%), $(C_2F_5)_2CO$ (74%), PhCHO (72%)EtCHO (71%), PhCOMe (82%)
R-M (product yield after acylation): *n*-BuLi (R'=Me, 71%, *E/Z*=72/28), PhLi (R'=Me, 64%, *E/Z*=77/23), PhMgBr (R'=Me, 48%, *E/Z*=83/17), Red-Al (R'=Ph, 55%, *E/Z*=43/57)

Figure 4. Reactions of *F*-enolate **1**.

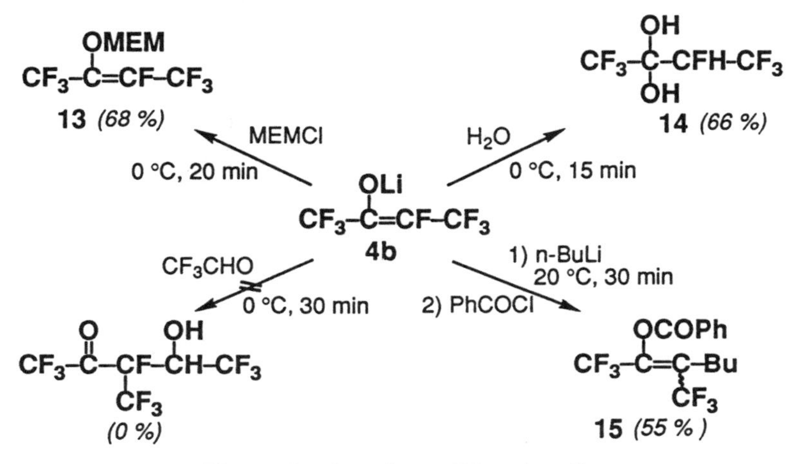

Figure 5. Reactions of *F*-enolate **4b**.

Table II. Electronic Properties of Free *F*-Enolates[a]

enolate	charge (e) on α-C	charge (e) on β-C	charge (e) on O	HOMO (eV)	LUMO (eV)
CH_3-C(O$^-$)=CH_2 (**16**) (α, β)	0.44	-0.82	-0.63	-1.83	11.60
CF_3-C(O$^-$)=CF_2 (**1**)	0.11	0.18	-0.61	-3.45	10.03
(*Z*)-CF_3-C(O$^-$)=$CFCF_3$ (**4b**)	0.21	-0.09	-0.53	-4.43	8.61
(*E*)-**4b**	0.19	-0.13	-0.53	-4.45	8.52

[a] Calculated at the optimized geometry obtained with a doublet-ζ basis set augmented by *d* functions on C and O. The basis set is from Dunning and Hay (Dunning, T. H; Hay, P. J. In *Methods of Electronic Structure Theory*; Schaefer, H. F. III, Ed.; Plenum Press: New York, 1977; Chapter 1) and has the form (9,5,1/9,5/4) / [3.2.1/3.2/2] in the order C,O/F/H.

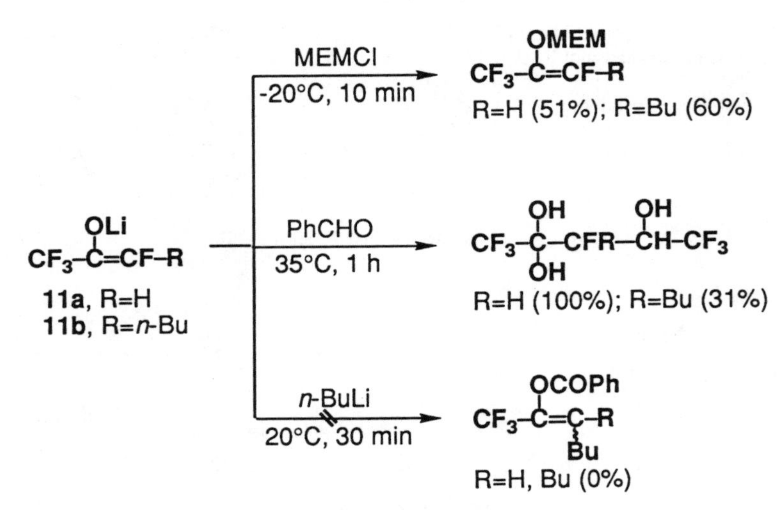

Figure 6. Reactions of enolate **11a** and **11b**.

CF_3 H O F

-0.59 e -0.26 e

(*Z*)-**11a**

LUMO: 9.98 eV

HOMO: -3.19 eV

Figure 7. Electronic properties of *(Z)*-**11a**.

substituted (**4b**) are well accommodated by the ab initio calculations on the free *F*-enolates, which have been done by Drs. Dixon and Smart (*4*).

The calculated electronic properties of the *F*-enolates and the hydrocarbon reference enolate **16** are summarized in Table II. The most revealing features are as follows. (a) The oxygen charge densities of the *F*-enolates and **16** are very similar. (b) The β-carbons of the *F*-enolates no longer have a large negative charge. (c) The HOMO and LOMO are both significantly lowered in **1** vs.**16** and also in **4b** vs. **1**. The β-C electrophilic reactivity of the *F*-enolates is rationalized by their relatively low lying LUMOs combined with the small positive or slightly negative charge on the β-carbon atoms. The lower LUMO level of **4b** relative to that of **1** nicely accounts for its enhanced *C*-electrophilicity, and its comparatively lower HOMO level is consistent with its poor reactivity in aldol reaction.

Reactivities of β-Hydro- and β-Alkyl-Substituted *F*-Enolates. The following scheme (Figure 6) shows the unique reactivity patterns observed with the β-hydro-*F*-enolate **11a** (R=H, M=Li) and the β-butyl-*F*-enolate **11b** (R=*n*-Bu, M=Li) (*7*). Both **11a** and **11b** shows a slightly higher *O*-nucleophilicity toward MEMCl and equally high aldol reactivity, compared with the parent *F*-enolate (**1**). Interestingly, however, both **11a** and **11b** no longer show the electrophilic reactivity toward *n*-BuLi.

Let's look at the calculated electronic properties of the free enolate (Z)-**11b** (*7*). As shown below (Figure 7), the β-carbon has a slightly larger negative charge, and the HOMO and LUMO levels are both slightly higher, compared with those of the parent *F*-enolate **1**. The equally high aldol reactivity of **11a** is rationalized by its comparably high lying HOMO, coupled with the large negative charge on the β-carbon. However, the rise of the LUMO level is apparently too small to account for its poor reactivity toward *n*-BuLi. Thus, we must await the additional calculations on the lithum coordinated *F*-enolate **11a**.

In summary, we have developed facile and regio- and stereoseleactive procedures to generate various types of *F*-enolates whose reactivities markedly differ from those of their hydrocarbon analogues but can be anticipated by *ab initio* molecular orbital calculations. This work convincingly reveals the interesting aspects of hitherto unexplored *F*-enolate chemistry, providing a new, basic technology for organofluorine synthesis. Further studies on the mechanism of *F*-enolate formation and synthetic utility are in progress. Finally, it should be noted that the ketone *F*-enolate chemistry described herein represents one extreme of enolate chemistry which we believe significantly promotes the understanding of the rich chemistry of partially fluorinated ketone enolates, a current subject of extensive investigations (*8-11*).

Acknowledgments. We are grateful to Drs. David A. Dixon and Bruce E. Smart of E. I. du Pont de Nemours & Co. Inc. for their excellent calculations and stimulating discussions, and Dr. Masamichi Maruta of Central Grass Co. for his helpful discussions. This work is partially supported by a Grant-in-Aid for Scientific Research from the Ministry of Education, Japan, the Chemical Materials Research & Development Fundation, Central Glass Co., and Du Pont, Japan, Ltd., which are gratefully acknowledged. We also thank Central Glass Co. and Asahi Glass Co. for the gift of HFIP and C_2F_5I, respectively.

Literature Cited.

(1) Bekker, R. A.; Melikyan, G. G.; Daytkin, B. L.; Knanyants, I. L. *Zh. Org. Khim.* **1976**, *12*, 1379.
(2) Farnharn, W. B.; Middleton, W. J.; Fultz, W. C.; Smart, B. E. *J. Am. Chem. Soc.* **1986**, *108*, 3125.

(3) Qian, C. -P.; Nakai, T. *Tetrahedron Lett.* **1988**, *29*, 4119
(4) Qian, C. -P.; Nakai, T.; Dixon, D. A.; Smart, B. E. *J. Am. Chem. Soc.* **1990**,
112, 4602.
(5) Chen, L. S.; Chen, G. J.; Tamborski, C. *J. Fluorine Chem.* **1984**, *26*, 341.
(6) Qian, C. -P.; Nakai, T. The 59th Annual Meeting of Chemical Society of Japan,
Yokohama, 1990, Abstr. 4D125.
(7) Qian, C. -P.; Maruta, M.; Nakai, T. The 56th Annual Meeting of Chemical Society of Japan, Tokyo, 1988, Abstr. 3XIA17.
(8) Kuroboshi, M.; Okada, Y.; Ishihara, T.; Ando, T. *Tetrahedron Lett.* **1987**, *28*,
350.
(9) Ishihara, T.; Yamaguchi, K.; Kuroboshi, M. *Chem. Lett.* **1989**, 1191.
(10) Whitten, J. P.; Barney, C. L.; Huber, E. W.; Bey, P.; McCarthy, J. R. *Tetrahedron Lett.* **1989**, *28*, 3649.
(11) Zeifman, Y. V.; Postovoi, S. A.; Vol'pin, I. M.; German, L. S. Proceedings of the 6th Regular Meeting of Soviet-Japanese Fluorine Chemists, Novosibirsk, 1989, I-1, and references cited therein.

RECEIVED October 19, 1990

Chapter 7

New Approaches to α-Fluoro and α,α-Difluoro Functionalized Esters

D. J. Burton, A. Thenappan, and Z-Y. Yang

Department of Chemistry, University of Iowa, Iowa City, IA 52242

Esters with one or two fluorines at the α-carbon are useful building blocks for construction of interesting and novel biologically active substrates. Alkylation of α-fluorocarboethoxy phosphonium ylides followed by hydrolysis of the resultant phosphonium salt with 5% aqueous sodium bicarbonate provides a useful preparative route to α-fluoroesters. Similarly, acylation/hydrolysis of either α-fluoro phosphonium ylides or α-fluorophosphonate anions gives a general route to 2-fluoro-3-oxo-esters. The α,α-difluoroesters can be prepared by Cu° catalyzed addition of iododifluoroacetates to olefins followed by reduction of the iodo addition adduct. Both terminal and internal olefins participate equally well in the addition reaction.

Elucidation of the mechanism of toxicity of fluoroacetate in living organisms led to increased interest into the preparation and properties of α-fluoroesters. More recently, the use of fluorine substituted esters as analytical probes and diagnostic tools in metabolic processes has added to their stature as important compounds in biochemistry*(1)*. In addition, α-fluoroesters have served as useful building blocks to more complex and interesting biological substrates.

A variety of methods have been utilized to incorporate a fluorine atom adjacent to an ester functionality, such as: (a) halogen exchange*(2-12)*; (b) reaction of diazonium intermediates (from α-amino acids) with Pyridine•$(HF)_n$*(13)*; (c) reaction of α-hydroxy esters with DAST or FAR*(14-16)* or the use of α-siloxyesters with fluorophosphoranes*(17)*; (d) condensation processes via the enolate of ethyl fluoroacetate*(18-23)*; (e) Reformatsky reaction with bromofluoroacetates*(24)*; and (f) electrophilic fluorination of carbanions or carbanion derivatives with $FClO_3$*(25,26)*, CF_3OF*(27-30)*, F_2*(31)*, CH_3COOF*(32,33)*, or N-F reagents*(34,35)*. However, limitations associated with these reported methods restrict their practicality. Consequently, we have explored the utility of α–fluorocarboalkoxy phosphorus ylides as a simple direct entry for the introduction of the -CFHCOOR functionality.

0097–6156/91/0456–0091$06.00/0

Preparation of 2-Fluoroalkanoates Via Alkylation-Hydrolysis of Ylides

Both α-fluorocarboalkoxy phosphonates*(36,37)* and α-fluorocarboalkoxy phosphonium salts*(38)* can be readily prepared from the commercially available ethyl bromofluoroacetate, or this ester can be readily obtained from the commercially available bromotrifluoroethene by application of the Organic Syntheses*(39)* preparation of ethyl chlorofluoroacetate. Subsequent deprotonation of either (**1**) or (**2**)

$$ROH + CF_2{=}CFBr \xrightarrow{RO} ROCF_2CFHBr \xrightarrow{H_2SO_4} CFHBrCOOR \quad (1)$$

$$(RO)_3P + CFHBrCO_2Et \xrightarrow{heat} \underset{(\mathbf{1})}{(RO)_2P(O)CFHCO_2Et} + RBr \quad (2)$$

R = Me,Et,i-Pr

$$R_3P + CFHBrCO_2Et \xrightarrow[RT]{solvent} \underset{(\mathbf{2})}{[R_3\overset{+}{P}CFHCO_2Et]Br^-} \quad (3)$$

R = C_6H_5, solvent = CH_2Cl_2

R = n-Bu, solvent = THF

gives the α-fluorophosphonate anion (**3**) and the α–fluorophosphonium ylide (**4**), respectively. The hydrocarbon analogs of (**3**) and (**4**) are powerful nucleophiles and have been demonstrated to react with a variety of alkylating agents such as alkyl halides and trialkyloxonium salts*(40-44)*. Thus, we anticipated that alkylation of either (**3**) or (**4**), followed by hydrolysis, would provide a direct synthesis of α-fluoroesters and that the particular analog produced could be varied via the proper choice of the alkylating agent.

Alkylation-Hydrolysis of α-Fluorophosphonate Carbanions. Our initial choice in the proposed scheme was the phosphonate analog, since we expected this carbanion to be more nucleophilic than the phosphonium analog and therefore more applicable to a variety of alkylating types. An initial concern in the reaction of (**3**)

$$(RO)_2P(O)CFHCOOEt \xrightarrow[\text{THF, -78°C}]{\text{BuLi}} [(RO)_2P(O)CFCOOEt]^-Li^+ + BuH \quad (3)$$

R = Et, i-Pr

$$\downarrow R'X \qquad (4)$$

$$(RO)_2P(O)CFR'COOEt + LiX \quad (5)$$

$$\downarrow \text{Hydrolysis}$$

$$R'CFHCOOEt$$

was the site of substitution. When (**3**) was treated with bromotrimethylsilane, only the O-silylated product was obtained (Thenappan, A., University of Iowa, unpublished data). Presumably, the regiochemistry here is controlled by the formation of the strong oxygen-silicon bond. When (**3**) is alkylated, however,

$$(EtO)_2P(O)\bar{C}FCOOEt + Me_3SiBr \longrightarrow (EtO)_2P(O)(F)C{=}C(OSiMe_3)(OEt) \qquad (5)$$

$$\downarrow H_2O$$

$$(EtO)_2P(O)CFHCOOEt$$

only the carbon alkylated product was observed (Table I).

Table I. **Alkylation of $[(RO)_2P(O)CFCOOEt]^-Li^+$**

$$[(RO)_2P(O)CFCOOEt]^-Li^+ + R'X \xrightarrow[\text{to reflux}]{-78°C} (RO)_2P(O)CFR'COOEt + LiX$$

X	R	R'	Yield (%)
I	Et	CH_3	69[a]
Br	Et	$C_6H_5CH_2$	86[a]
Br	Et	CH_2=$CHCH_2$	56[a]
Br	i-Pr	$CH_3CHC_6H_5$	60[b]
I	i-Pr	CH_3CHCH_3	72[b]

[a]Mixture of phosphonate and mono-dealkylated phosphonate, ^{19}F NMR yield;
[b]isolated yield

The alkylation reaction occurred easily with both primary and secondary halides. When R = Et in the phosphonate carbanion, the alkylated phosphonate suffers partial dealkylation to $O^-(RO)P(O)CFR'COOEt$ by SN2 attack of the lithium halide produced in the reaction. This side-reaction can be easily suppressed by use of the corresponding isopropyl phosphonate carbanion. Thus, this straightforward alkylation of (**3**) appeared promising as an entry to α-fluoroesters. Alkylation occurred only at carbon and the absence of hydrogen at the α-carbon in the phosphonate precluded any transylidation process, thus allowing total utility of the phosphonate carbanion only in the desired alkylation reaction without concomitant loss of the phosphonate carbanion in acid-base side-reactions.

Unfortunately, reasonable attempts to complete the reaction sequence via hydrolysis of (**5**) failed. The use of 5% sodium bicarbonate, 5% sodium carbonate

$$(RO)_2P(O)CFR'COOEt \xrightarrow{\text{Hydrolysis (}\times\text{)}} R'CFHCOOEt \quad (6)$$

and 5% sodium hydroxide at both room temperature and reflux failed to cleave (**5**). Reductive cleavage of (**5**) with zinc and acetic acid at reflux also failed. Treatment of (**5**) with moist potassium fluoride did slowly cleave (**5**) to give the α-fluoroester. However, the formation of the toxic $(RO)_2P(O)F$ via this mode of cleavage was not attractive for large scale reactions and prompted us to explore an alternative ylide route to these esters.

Alkylation-Hydrolysis of α-Fluorophosphonium Ylides. The ease of alkylation of (**3**) suggested that alkylation of the less nucleophilic phosphonium analog (**4**) could also be accomplished. When (**4**) was generated from the phosphonium salt (**2**), the ^{19}F NMR spectrum of the resulting solution exhibited two doublets in a 1:1 ratio, consistent with the formation of the two geometrical isomers

$$[Bu_3\overset{+}{P}CFHCOOEt]Br^- \xrightarrow[\text{THF, -78°C}]{\text{BuLi}} [Bu_3\overset{+}{P}\overset{-}{C}FCOOEt] + BuH + LiBr \quad (7)$$

(**4**) ↔ (**4a**): $[Bu_3P(F)C{=}C(O^-)(OEt)]$

of (**4a**). The ready detection of (**4a**) gave some concern as to the regiospecificity of a subsequent alkylation process. However, alkylation occurred only at carbon; no O-alkylation was detected. The alkylation at carbon and/or oxygen are readily distinguishable by virtue of the differences in the multiplicity of the fluorine signal

$$(\mathbf{4a}) \xrightarrow{R'X} [Bu_3\overset{+}{P}CFR'COOEt]X^- \ (\mathbf{6}) \quad (8)$$

$$(\mathbf{4a}) \overset{R'X}{\nrightarrow} [Bu_3\overset{+}{P}(F)C{=}C(OR')(OEt)]X^- \ (\mathbf{7})$$

in the ^{19}F NMR spectrum. For example, when R' = CH_3, (**6**) exhibits a doublet of quartets, whereas if (**7**) were formed, only a doublet would be observed.

As one might expect, the nucleophilicity of (**4**) is less than (**3**), and this difference in reactivity is reflected in the alkylation reaction. (**4**) reacts readily with activated bromides (allylic, benzylic) and the reactions are complete within 12 hours; activated chlorides did not react. With primary alkyl iodides the reaction is slow (12-120 hours depending on the chain length). With secondary halides, such as 2-bromopropane, ethyl-2-bromopropionate, (1-bromoethyl)benzene and 2-iodopropane, no alkylation was observed.

When the alkylating agent is a substituted allylic bromide, reaction could occur at either the α- or the γ-carbon. Only α-attack is observed and the phosphonium salt (**8**) from regiospecific alkylation at the less hindered carbon is produced. Again, the two possible regioisomers are distinguishable by virtue of the difference in the multiplicity of their fluorine signals in the ^{19}F NMR spectrum. Salt (**8**) appears as a doublet of doublet of doublets, whereas (**9**) would be expected to be a doublet of doublets.

$$Bu_3\overset{+}{P}\overset{-}{C}FCOOEt + RCH{=}CHCH_2Br \xrightarrow{\alpha\text{-attack}} [Bu_3\overset{+}{P}CF(COOEt)CH_2CH{=}CHR]Br^- \ (\mathbf{8}) \tag{9}$$

$$\xrightarrow{\gamma\text{-attack}\ \times} [Bu_3\overset{+}{P}CF(COOEt)CHRCH{=}CH_2]Br^- \ (\mathbf{9})$$

In contrast to the difficulty experienced in the hydrolysis of (**5**), the phosphonium salts (**6**) were all easily hydrolyzed by 5% sodium bicarbonate, 5% sodium carbonate or 5% sodium hydroxide to give the α-fluoroester. These results are summarized in Table II. The increased positive charge on phosphorus in(**6**) compared to (**5**) facilitates attack by the hydrolysis reagent and allows the overall scheme to succeed. Thus, α-fluoroesters can be readily prepared via alkylation and hydrolysis of (**4**). The reaction is not as general with (**4**) (secondary halides did not react), but for primary and activated halides, the reaction does provide a convenient, facile, easily scaled up synthesis for a variety of α-fluoroesters.

Preparation of 2-Fluoro-3-Oxoalkanoates Via Acylation-Hydrolysis of Ylides

The importance of α-fluoro-β-ketoesters stems from their use as synthons in the preparation of biologically active monofluorinated heterocycles*(45)* and fluorinated isoprenyl derivatives *(46)* which have found application as hyperlipidemic drugs *(47)*, as hormone substitutes *(48)*, and in cancer chemotherapy *(49)*.

Several routes to this class of compounds have been reported, such as: (a) crossed Claisen condensation reactions *(50-53)*; (b) acylation of the anion derived from ethyl fluoroacetate *(54)* or self-condensation of the anion derived from ethyl bromofluoroacetate *(55)*; (c) electrophilic fluorination of the anion of β-ketoesters *(56,57)*; (d) acylation-hydrolysis of fluoroolefins *(58)*; and (e) acylation of fluorine-containing ketene silyl acetals (Easdon, J.C., University of Iowa, unpublished data). The limitations associated with these methods and the success achieved in the alkylation-hydrolysis of α-fluoro phosphorus ylides prompted us to examine acylation-hydrolysis of these α-fluoro ylides as a general route to 2-fluoro-3-oxoesters.

The acylation of ylides has been utilized as a method for the elaboration of a simple ylide to a functionalized ylide. Transylidation of the acylated ylide gives the

$$R_3\overset{+}{P}\overset{-}{C}HR' + R''C(O)Cl \longrightarrow [R_3\overset{+}{P}CHR'C(O)R'']Cl^- \xrightarrow{R_3\overset{+}{P}\overset{-}{C}HR'} R_3\overset{+}{P}\overset{-}{C}R'C(O)R'' + [R_3\overset{+}{P}CH_2R']Cl^- \tag{10}$$

Table II. Preparation of $RCH_2CFHCOOEt$

$$Bu_3P{=}CFCOOEt \xrightarrow[\text{1) } RCH_2X \text{ 2) } NaHCO_3\text{ (aq)}]{THF/-78^\circ} RCH_2CFHCOOEt$$

R	Yield[a]	bp (°C)/mm Hg
H	59 (75)	50/64
CH_3	42 (87)	60-1/60
CH_3CH_2	42 (74)	94-5/117
$CH_3(CH_2)_2$	34 (61)	93-4/63
$CH_3(CH_2)_5$	52 (72)	93-4/5
$CH_3(CH_2)_8$	44 (50)	115-6/1
CH_2=CH	52 (73)	82-3/28
Ph	59 (61)	69-70/0.3
Ph-CH=CH(E+Z)	45 (76)	112-3/0.5
CH_3-CH=CH(E+Z)	38 (72)	84-5/25
CH_3O-C(O)-CH=CH(E)	41 (45)	72-80/0.6

a) Isolated yields are based on ethyl bromofluoroacetate, and the yields in parentheses are determined by ^{19}F NMR analysis.

new functionalized ylide. With α-fluoro phosphorus ylides, the transylidation process is precluded, thus the acylation reaction could potentially be stopped cleanly after the initial acylation step.

Thus, with model substrates, we initially explored both the acylation of (**3**) and (**4**). Both ylides readily underwent acylation with acyl halides or acetic anhydride to give the C-acylated phosphonate or phosphonium salt.

Method [A]:

$$(\mathbf{4}) + RC(O)Cl \longrightarrow [Bu_3\overset{+}{P}CF(COOEt)C(O)R]Cl^- \; (\mathbf{10}) \xrightarrow{NaHCO_3,\ aq.} RC(O)CFHCOOEt \qquad (11)$$

Method [B]:

$$(\mathbf{3}) + R'C(O)Cl \longrightarrow (RO)_2P(O)CF(COOEt)C(O)R' \; (\mathbf{11}) \xrightarrow{NaHCO_3,\ aq.} R'C(O)CFHCOOEt \qquad (12)$$

As noted earlier in the alkylation study of α-fluoro phosphorus ylides, hydrolysis of the resultant α-fluoro phosphonium salts and α-fluorophosphonates do not necessarily parallel each other. This difference is again exhibited in the hydrolysis of the acylated products of the α-fluoro ylides. For example, the acylation product (**10**) is readily hydrolyzed at room temperature to give the desired α-fluoro-β-ketoester. The results of the acylation-hydrolysis of (**4**) with a variety of alkyl, cycloalkyl and aryl acyl halides are summarized in Table III .

In contrast to the straightforward facile acylation-hydrolysis reaction of the α-fluoro phosphonium ylide, the acylated product from the α-fluoro phosphonate carbanion is cleaved by base in two different ways. When R is a hydrocarbon group, such as CH_3 or $C_6H_5CH_2$, attack at the acyl carbon with bases, such as sodium bicarbonate, sodium carbonate, sodium hydroxide, and potassium silanoate is favored (Path II in equation 13) with resultant elimination of the α-fluorophosphonate anion. Less than 10% of the desired 2-fluoro-3-oxoester is observed. However, when R is a halofluoroalkyl group (CF_3, CF_2Cl, C_3F_7), attack of the base (aqueous sodium bicarbonate) occurs only at phosphorus (Path I in equation 13) and the 2-fluoro-3-

Table III. Preparation of RC(O)CFHCOOEt

$$Bu_3P{=}CFCOOEt \xrightarrow[\text{2) NaHCO}_3\text{ (aq)}]{\text{1) RC(O)Cl}} RC(O)CFHCOOEt \text{ (Method A)}$$

$$(EtO)_2P(O)CFHC(O)OEt \xrightarrow[\text{2) R}_f\text{C(O)Cl; 3) NaHCO}_3\text{ (aq)}]{\text{1) n-BuLi}} R_fC(O)CFHCOOEt \text{ (Method B)}$$

R	Method	Isolated[a] Yield (%)	bp (°C)/ mmHg
CH_3[b]	A	60	60-62/4.0
CH_3CH_2	A	50	48-52/2.2
$(CH_3)_2CH$	A	58	62-63/2.5
$(CH_3)_3C$	A	56	56-61/2.0
C_6H_{11}	A	68	76-84/0.5
EtO	A	50	72-74/3.5
C_6H_5	A	70	112-114/0.4
$CH_3OCOCH_2CH_2$	A	38	92-97/0.45
EtS	B	57	74-77/0.5
n-C_3F_7	B	77	45-47/5.0
CF_3	B	60	42-43/43
CF_2Cl	B	67	60-65/38

a) Isolated yields are based on acid chloride.

b) Acylation using acetic anhydride.

SOURCE: Reproduced with permission from reference 78. Copyright 1989 Pergamon.

$$(EtO)_2P(O) \overset{I}{-\wr-} \underset{|\atop COOEt}{CF} \overset{II}{-\wr-} C(O)R + \text{base} \longrightarrow \begin{cases} \xrightarrow{I} RC(O)CFHCOOEt \\ \xrightarrow{II} (EtO)_2P(O)CFHCOOEt \end{cases} \quad (13)$$

oxoester (as the hydrate) is formed in good yield (isolated via distillation from sulfuric acid). In this case, the acyl carbonyl group is presumably hydrated; hence attack at the acyl carbon is inhibited. Thus, attack at the phosphoryl group is favored with elimination of the α-fluoro-β-ketoester anion.

Therefore, when R is a hydrocarbon group, the acylation-hydrolysis of (**4**) is the preferred route. When R is perfluoroalkyl or halofluoroalkyl, the acylation-hydrolysis of (**3**) is the preferred route. Thus, by proper choice of the α-fluoro phosphorus ylide, the acylation-hydrolysis methodology provide a facile entry to a wide variety of α-fluoro-β-ketoesters from readily available precursors.

Preparation of α,α-Difluoroesters

Bioactive compounds that contain the difluoromethylene group adjacent to the carbonyl functionality have been the subject of increased research efforts in recent years*(59)*. The most widely utilized methods that have been employed to introduce this group into organic molecules have been: (a) Reformatsky reaction of halodifluoroacetates *(60-66)*; (b) elaboration of difluoroketene silyl acetals *(67-69)*; (c) metal catalyzed addition of 3-bromo-3,3-difluoropropene to aldehydes and ketones *(70)*; and (d) alkylation of $CuCF_2COOR$*(71)*. The modest yields associated with these methods prompted us to explore alternative methodology for the preparation of this useful building block.

Based on previous reports by Coe*(72)* and Chen *(73-75)*, we anticipated that copper catalyzed additions of iododifluoroacetates to olefins could be accomplished under mild conditions. Subsequent reduction of the addition adduct would give the α,α-difluoroesters.

$$RCH{=}CHR' + ICF_2CO_2R'' \xrightarrow[50\text{-}60°C]{Cu°} RCHICH(R')CF_2CO_2R'' \xrightarrow{\text{Reduction}} RCH_2CH(R')CF_2CO_2R'' \quad (14)$$

Recently, a variant of this type of process was reported by Barth and O-Yang*(76)* who reported the free-radical cyclization of unsaturated α-fluoro-α-iodoesters and amides.

The initial focus of our work was to test the generality of the copper catalyzed addition reaction. We found that under mild conditions (50-60°C, no solvent) in the presence of 10-20 mol % copper powder, iododifluoroacetates added cleanly to

Table IV. Preparation of $R^1CHICH(R^2)CF_2CO_2R$

$$ICF_2CO_2R + R^1CH{=}CHR^2 \xrightarrow[50\text{-}60°C]{Cu} R^1CHICH(R^2)CF_2CO_2R$$

R	R^1	R^2	Product	Yield
Me	H	n-Bu	n-BuCHI$CH_2CF_2CO_2$Me	75
Me	H	n-C_5H_{11}	n-C_5H_{11}CHI$CH_2CF_2CO_2$Me	65
Me	H	$SiMe_3$	Me_3SiCHI$CH_2CF_2CO_2$Me	83
Me	-$(CH_2)_4$-		2-iodocyclohexyl-CF_2CO_2Me (ring structure: I, CF_2CO_2Me)	78
Et	H	n-Bu	n-BuCHI$CH_2CF_2CO_2$Et	65
Et	H	n-C_5H_{11}	n-C_5H_{11}CHI$CH_2CF_2CO_2$Et	76
Et	H	Me_3Si	Me_3SiCHI$CH_2CF_2CO_2$Et	70
Et	n-C_3H_7	n-C_3H_7	n-C_3H_7CHICH(n-C_3H_7)CF_2CO_2Et	72
Et	-$(CH_2)_4$-		2-iodocyclohexyl-CF_2CO_2Et (ring structure: I, CF_2CO_2Et)	75
^{i}Pr	H	Me_3Si	Me_3SiCHI$CH_2CF_2CO_2{}^i$Pr	72
^{i}Pr	H	n-C_6H_{13}	n-C_6H_{13}CHI$CH_2CF_2CO_2{}^i$Pr	76
^{i}Pr	H	n-C_6H_{13}	n-C_6H_{13}CHI$CH_2CF_2CO_2{}^i$Pr	80[a]
^{i}Pr	H	n-C_6H_{13}	n-C_6H_{13}CHI$CH_2CF_2CO_2{}^i$Pr	73[b]

a) in benzene; b) in hexane

SOURCE: Reproduced with permission from reference 79. Copyright 1989 Elsevier.

olefins to give the 1:1 addition adduct in high yields. In non-coordinating solvents, such as hexane or benzene, similar results were obtained.

Terminal, internal and cyclic alkenes gave similar results. Vinyl silanes also worked well. The results of this copper catalyzed addition reaction are summarized in Table IV.

Preliminary work with tri-n-butyl tin hydride and other reducing reagents has demonstrated that the addition product can be readily converted to the α,α-difluoroester.

Thus, this facile addition-reduction sequence provides a convenient entry to the α,α-difluoroester building block and will facilitate the use of these precursors for elaboration to interesting biologically active materials.

Acknowledgments. We thank the National Science Foundation and the Air Force Office of Scientific Research for support of this work.

Literature Cited

(1) *Fluorine-Containing Molecules: Structure, Reactivity, Synthesis, and Applications*; Liebman, J.F.; Greenberg, A.; Dolbier, W.R., Eds.; VCH Publishers: New York, N.Y., 1988; Chapter 11.
(2) Swarts, F. *Bull. Acad. Roy. Belg.* **1896**, *31*, 675.
(3) Swarts, F. *Bull. Acad. Roy. Belg.* **1896**, *15*, 1134.
(4) Saunders, B.C.; Stacey, G.J. *J. Chem. Soc.* **1948**, 1773.
(5) Bergmann, E.D.; Blank, I. *J. Chem. Soc.* **1953**, 3786.
(6) Kim, Y.S.; Kim, K.S. *Chim. Abstr.* 1969, *71*, 123753 w; *Daehan Hwahak Hwoejee* **1969**, *13*, 68.
(7) Normant, J.F.; Bernardin, J. *C.R. Acad. Sci. Ser. C* **1969** *268*, 2352.
(8) Birdsall, N.J.M. *Tetrahedron Lett.* **1971**, *28*, 2675.
(9) Kobayashi, Y.; Taguchi, T.; Terada, T.; Oshida, J.; Morisaki, M.; Ikekawa, N. *J. Chem. Soc., Perkin Trans. 1* **1982**, 85.
(10) Ishikawa, N.; Kitazume, T.; Yamazaki, T.; Mochida, Y.; Tatsuno, T. *Chem. Lett.* **1981**, 761.
(11) Colonna, S.; Gelbard, G.; Re, A.; Cesarotti, E. *J. Chem. Soc., Perkin Trans. 1* **1979**, 2248.
(12) DeKleijn, J.P.; Seetz, J.W.; Zawierko, J.F.; Van Zanten, B. *Int. J. Appl. Radiat. Isot.* **1977**, *28*, 591.
(13) Olah, G.A.; Welch, J.T.; Yankar, Y.D.; Nojima, M.; Kerekes, I.; Olah, J.A. *J. Org. Chem.* **1979**, *44*, 3872.
(14) Middleton, W.J. *J. Org. Chem.* **1975**, *40*, 574.
(15) Lowe, G.; Potter, B.V.L. *J. Chem. Soc., Perkin Trans. 1* **1980**, 2029.
(16) Takaoka, A.; Iwakiri, H.; Ishikawa, N. *Bull. Chem. Soc. Jpn.* **1979**, *52*, 3377.
(17) Costa, D.J.; Boutin, N.E.; Riess, J.G. *Tetrahedron* **1974**, *30*, 3793.
(18) Bergmann, E.D.; Szinai, S. *J. Chem. Soc.* **1956**, 1521.
(19) Fraisse, J.R.; Nguyen, T.L. *Bull. Soc. Chim. Fr.* **1967**, 3904.
(20) Baerwolf, D.; Reffschlaeger, J.; Langen, P. *Nucleic Acids Symp. Ser.*,**1981**, *9*, 45.
(21) Blank, I.; Mager, J.; Bergmann, E.D. *J. Chem. Soc.*, **1955**, 2190.
(22) Bobek, M.; Farkas, J.;' Gut, J. *Czech. Chem. Commun.* **1967**, *32*, 1295.
(23) Elkik, E.; Parlier, A.; Dahan, R. *C.R. Acad. Sci. Ser.* **1975** *C 281*, 337.
(24) Brandange, S.; Dahlman, O.; Morch, L. *J. Am. Chem. Soc.* **1981**, *103*, 4453.
(25) Alekseeva, L.V.; Lundin, B.N.; Burde, N.L. *Zh. Obshch. Khim.* **1967**, *37*, 1754.

(26) Gershon, H.; Schulman, S.G.; Spevack, A.D. *J. Med. Chem.* **1967**, *10*, 536.
(27) Barton, D.H.R.; Godinho, L.S.; Hesse, R.H.; Pechet, M.M. *J. Chem. Soc., Chem. Commun.* **1968**, 804.
(28) Barton, D.H.R.; Ganguly, A.K.; Hesse, R.H.; Loo, S.N.; Pechet, M.M. *J. Chem. Soc., Chem, Commun.* **1968**, 806.
(29) Barton, D.H.R.; Danks, L.J.; Ganguly, A.K.; Hesse, R.H.; Tarzia, G.; Pechet, M.M. *J. Chem. Soc., Chem. Commun.* **1969**, 227.
(30) Middleton, W.J.; Bingham, E.M. *J. Am. Chem. Soc.* **1980**, *102*, 4845.
(31) Tsushima, T.; Kawada, K.; Tsuji, T.; Misaki, S. *J. Org. Chem.* **1982**, *47*, 1107.
(32) Rozen, S.; Lerman, O.; Kol, M. J. Chem. Soc., *J. Chem. Soc., Chem. Commun.* **1981**, 443.
(33) Rozen, S.; Lerman, O.; Kol, M.; Hebl, D. *J. Org. Chem.* **1985**, *50*, 4753.
(34) Purington, S.T.; Jones, W.A. *J. Org. Chem.* **1983**, *48*, 761.
(35) Barnette, W.E. *J. Am. Chem. Soc.* **1984** *106*, 452.
(36) Machleidt, H.; Wessendorf, R. *Lieg. Ann.* **1964**, *674*, 1.
(37) Elkik, E.; Oudotte, M.I.; Normant, H. *C.R. Acad. Sci. Ser. II* **1981**, *292*, 1023.
(38) Gurusamy, N.; Burton, D.J. 188th ACS National Meeting, Philadelphia, Pa. 1984, Abstract FLUO 12.
(39) Englund, B. *Organic Syntheses Collective Volume*; Rabjohn, N. Ed.; John Wiley and Sons: New York, N.Y., 1963; Vol. IV, pp 423-426.
(40) Johnson, A.W. *Ylid Chemistry*; Academic Press: New York, N.Y.; 1966.
(41) Cadogan, J.I.G. *Organophosphorus Reagents in Organic Synthesis*; Academic Press: London, 1979.
(42) Arbuzov, A.E.; Razumov, A.I. *Chem. Abst.* **1929**, *23*, 4444; *J. Russ. Phys. Chem.* **1924**, *61*, 623.
(43) Bestmann, H.J.; Schulz, H. *Tetrahedron Lett.* **1960**, *4*, 5.
(44) Markl, G. *Tetrahedron Lett.* **1962**, 1027.
(45) Bergmann, E.D.; Cohen, S.; Shahak, I. *J. Chem. Soc.* **1959**, 3278.
(46) Montellano, P.R.O.; Vinson, W.A. *J. Org. Chem.* **1977**, *42*, 2013.
(47) Machleidt, H. , U.S. Patent 3 277 147, Oct. 4, 1966.
(48) Camps, F.; Coll, J.; Messeguer, A.; Roca, A. *Tetrahedron Lett.* **1976**, *10*, 791.
(49) Sporn, M.B.; Dunlop, N.M.; Newton, D.L.; Smith, J.M. *Fed. Proc., Fed. Am. Soc. Exp. Biol.* **1976**, *35*, 1332.
(50) Swarts, F. *Bull. Class Sci. Acad. Roy. Belg.***1926**, *12*, 692.
(51) Henne, A.L.; Newman, M.S.; Quill, L.L.; Staniforth, R.A.*J. Am. Chem. Soc.* **1947**, *69*, 1819.
(52) McBee, E.T.; Pierce, O.R.; Kilbourne, H.W.; Wilson, E.R. *J. Am.Chem. Soc.* **1953**, *75*, 3152.
(53) Elkik, E.; Oudptte, M. *Compt. Rend.* **1972**, *C 274*, 1579.
(54) Bergmann, E.D.; Cohen, S.; Shahak, I. *J. Chem. Soc.* **1959**, 3278.
(55) Elkik, I.; Le Bianc, M.; Hamid, Assadi-Far *Compt. Rend.* **1971**, *C 272*, 1895.
(56) Yemul, S.S.; Kagan, H.B. *Tetrahedron Lett.* **1980**, *21*, 277.
(57) Pattison, F.L.M.; Peters, D.A.V.; Dean, F.H. *Can. J. Chem.* **1965**, *43*, 1689.
(58) Ishikawa, N.; Takaoka, A.; Iwakiri, H.; Kubota, S.; Kagaruki, S.R.F. *Chem. Lett.* **1980**, 1107.
(59) Welch, J.T. *Tetrahedron* **1987**, *43*, 3123.
(60) Hallinan, E.A.; Fried, J. *Tetrahedron Lett.* **1984**, 2301.

(61) Fried, J.; Hallinan, E.A.; Szwedo, M.J. *J. Am. Chem. Soc.* **1984**,*106*, 3871.
(62) Gelb, M.H.; Svaren, J.P.; Abeles, R.H. *Biochemistry* **1985**, *24*, 1813.
(63) Thaisrivongs, S.; Pals, D.T.; Kati, W.M.; Turner, S.R.; Thomasco, L.M. *J. Med. Chem.* **1985**, *28*, 1553.
(64) Thaisrivongs, S.; Pals, D.T.; Kati, W.M.; Turner, S.R.; Thomasco, L.M.; Watt, W. *J. Med. Chem.* **1986**,*29*, 2080.
(65) Burton, D.J.; Easdon, J.C. *J. Fluorine Chem.* **1988**, *38*, 125.
(66) Lang, R.W.; Schuab, B. *Tetrahedron Lett.* **1988**, 2943.
(67) Burton, D.J.; Easdon, J.C., 190th National Meeting of the American Chemical Society, Chicago, Il., September 1985, Abst. FLUO 016.
(68) Burton, D.J.; Easdon, J.C., 12th International Symposium on Fluorine Chemistry, Santa Cruz, Ca., August 1988, Abst. #110.
(69) Kitagawa, O.; Taguchi, T.; Kobayashi, Y. *Tetrahedron Lett.* **1988**, 1803.
(70) Yang, Z-Y.; Burton, D.J. *J. Fluorine Chem.* **1989**, *44*, 339.
(71) Taguchi, T.; Kitagawa, O.; Morikawa, T.; Nishiwaki, T.; Uehara, H.; Endo, H.; Kobayashi, Y. *Tetrahedron Lett.* **1986**, 6103.
(72) Coe, P.L.; Milner, N.E. *J. Organometal. Chem.* **1972**, *39*, 395.
(73) Chen, Q.Y.; Yang, Z-Y. *J. Fluorine Chem.* **1985**, *28*, 399.
(74) Chen, Q.Y.; Yang, Z-Y. *Acta Chimica Sinica* **1986**, *44*, 265.
(75) Chen, Q.Y.; Yang, Z-Y.; He, Y.B. *J. Fluorine Chem.* **1987**, *37*, 171.
(76) Barth, F.; O-Yang, C., *Tetrahedron Lett.* **1990**, 1121.
(77) Thenappan, A.; Burton, D.J. *Tetrahedron Lett.* **1989**, 3641.
(78) Thenappan, A.; Burton, D.J. *Tetrahedron Lett.* **1989**, 6113.
(79) Yang, Z-Y.; Burton, D.J. *J. Fluorine Chem.* **1989**, *45*, 435.

RECEIVED August 17, 1990

Chapter 8

Terminal Fluoroolefins

Synthesis and Application to Mechanism-Based Enzyme Inhibition

Philippe Bey, James R. McCarthy, and Ian A. McDonald

Merrell Dow Research Institute, 2110 East Galbraith Road, Cincinnati, OH 45215

The incorporation of a fluoroolefin functionality into a substrate of a particular enzyme is often an effective way to design a mechanism-based inhibitor of an enzyme that catalyzes an oxidation step during turnover of substrate to product. A review of published synthetic methods leading to fluoroolefins is presented with particular focus on examples relevant to enzyme inhibition.

Over the past 15 years, the concept of mechanism-based inhibition has been exploited successfully to develop a wealth of new enzyme inhibitors of therapeutic relevance *(1)*. The prototype of mechanism-based inhibitors is the suicide inhibitor. Suicide inhibitors are chemically unreactive pseudo-substrates of the target enzymes which incorporate in their structures a latent reactive group. Activation of the latent group during enzymatic turnover generates a species that eventually inactivates the target enzyme, usually through formation of a covalent bond with a residue of the active site of the enzyme or of the cofactor. The functionalities suitable as latent reactive groups obviously depend upon the mechanism of action of the target enzymes. For example, double bonds have proven useful in the design of mechanism-based inhibitors for enzymes that catalyze an oxidation step during turnover of substrates to products *(2)*. When the double bond is located on the carbon atom of the substrate next to the function that is oxidized, an electrophilic Michael acceptor is generated in the enzyme's active site, provided that the modified unsaturated substrate is still turned over by the target enzyme. The Michael acceptor can eventually alkylate an adventitious nucleophilic residue in the active site, resulting in the formation of a covalent adduct between the inhibitor and enzyme which leads to inactivation of the target enzyme. In principle, the chemical reactivity of the double bond can be manipulated by adding a fluorine atom on the distal carbon atom of the double bond; the

0097–6156/91/0456–0105$08.25/0

fluorine atom is similar to an hydrogen atom in terms of steric hindrance *(3)*. The fluorinated unsaturated derivative can, therefore, be expected to be a substrate for the target enzyme if the corresponding unsaturated derivative is also a substrate. The β-fluorinated Michael acceptor that would be formed is more electrophilic and consequently more reactive towards a nucleophilic residue than the non-fluorinated system. Moreover, the β-fluoro-α,β-unsaturated system can add the enzyme nucleophilic residue in a Michael-type addition-elimination reaction as indicated in Figure 1. A potential advantage is that the resulting covalent bond between the enzyme and the inhibitor would be more stable since a retro-Michael reaction is no longer possible.

In this chapter, we discuss the synthetic methodologies used to prepare fluoroolefins and present examples of mechanism-based inhibitors of amine oxidases, γ-aminobutyric acid transaminase and *S*-adenosylhomocysteine hydrolase which incorporate this structural functionality. We have restricted our discussion to the syntheses of terminal mono-, di- and trifluoroolefins, omitting the large body of synthetic endeavour directed towards other fluoro olefins *(4)*.

Wittig Reactions and Organometallic Approaches to Fluoroolefins

The Wittig reaction has served as a versatile method to mono- and difluoroolefins. Fuqua *(5,6)* and Burton *(7,8)* first applied this reaction to the synthesis of difluoroolefins (3) by the *in situ* generation of triphenyldifluoromethylenephosphorane (2) from sodium chlorodifluoroacetate (1) and triphenylphosphine in the presence of aldehydes *(7,8)* or ketones *(6)* (Scheme 1). When the reaction was carried out with an unsaturated carbonyl compound no difluorocyclopropanes were formed *(8)*, suggesting the absence a discrete difluorocarbene intermediate. At the time this method was developed there were no other simple general methods to difluoroolefins *(9,10)*. The synthesis of fluoromethyltriphenylphosphonium iodide (5) from fluoroiodomethane (4) provided a direct route to monofluoro-substituted olefins *(11,12)*. The scope of the reaction, as well as alternate methods for the synthesis of 5 and fluoromethylenetriphenylphosphorane (6), were studied by Burton and Greenlimb (13) (Scheme 2). Although improved yields of fluoroolefins 7 (from 5) were realized when potassium t-butoxide was used in addition to *n*-butyllithium, these were generally less than 50%. Moreover, extremely dry solvents were required to obtain reproducible yields and the availability and expense of the starting halofluoromethanes, CH_2FI and $CHFI_2$, limited the utility of the method.

Replacement of iodofluoromethane (4) in Scheme 2 with dibromodifluoromethane (9) provided a new route to the Wittig reagent $Ph_3P{=}CF_2$ (2) *(14)*. In contrast to 2 (obtained from 10), which appears best suited for reaction with aldehydes, (dimethylamino)difluoromethylenephosphorane (11) adds to both aldehydes and unactivated ketones (Scheme 3). The reaction conditions, however, are still extremely sensitive to moisture.

Difluoromethylenation of the activated ketone 12 was achieved chemoselectively with $Ph_3P/CF_2Br_2/Zn$ (Scheme 4); the lactone carbonyl was unreactive *(15)*. However, the reagent generated from $CF_2Br_2/(Me_2N)_3P$ reacted with a formate ester (14) in high yield (Scheme 5) *(16)* and the lactone (16) could be converted to 17 by

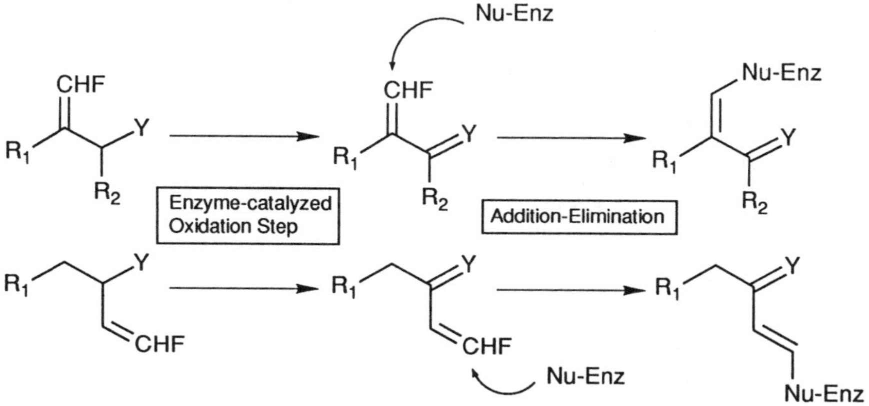

Nu-Enz

CHF

Y

R_1

R_2

Enzyme-catalyzed Oxidation Step

Addition-Elimination

CHF

Nu-Enz

Figure1. *Rational for the design of fluoroolefin-containing mechanism-based enzyme inhibitors*

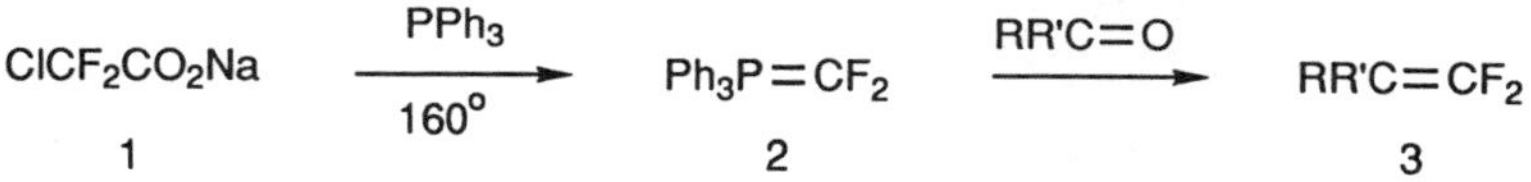

$$\underset{1}{ClCF_2CO_2Na} \xrightarrow[160^o]{PPh_3} \underset{2}{Ph_3P{=}CF_2} \xrightarrow{RR'C{=}O} \underset{3}{RR'C{=}CF_2}$$

Scheme 1

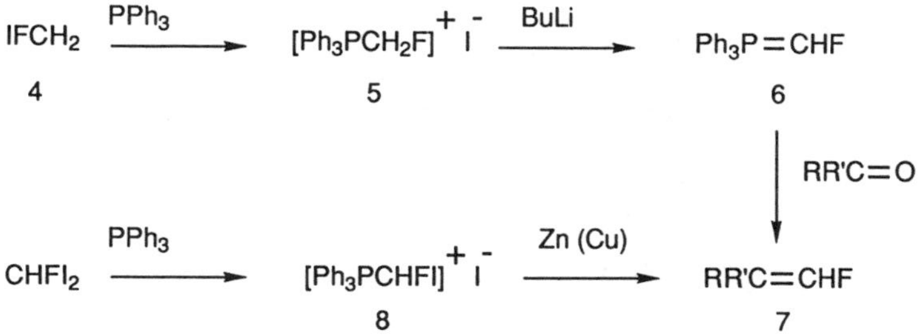

$$\underset{4}{IFCH_2} \xrightarrow{PPh_3} \underset{5}{[Ph_3PCH_2F]^+ I^-} \xrightarrow{BuLi} \underset{6}{Ph_3P{=}CHF}$$

$$\underset{6}{Ph_3P{=}CHF} \xrightarrow{RR'C{=}O} \underset{7}{RR'C{=}CHF}$$

$$CHFI_2 \xrightarrow{PPh_3} \underset{8}{[Ph_3PCHFI]^+ I^-} \xrightarrow{Zn\ (Cu)} \underset{7}{RR'C{=}CHF}$$

Scheme 2

$$\underset{11}{(Me_2N)_3P{=}CF_2} \xleftarrow{2\ (Me_2N)_3P} \underset{9}{CBr_2F_2} \xrightarrow{PPh_3} \underset{10}{[Ph_3PCBrF_2]^+ Br^-} \xrightarrow{Zn} \underset{2}{Ph_3P{=}CF_2}$$

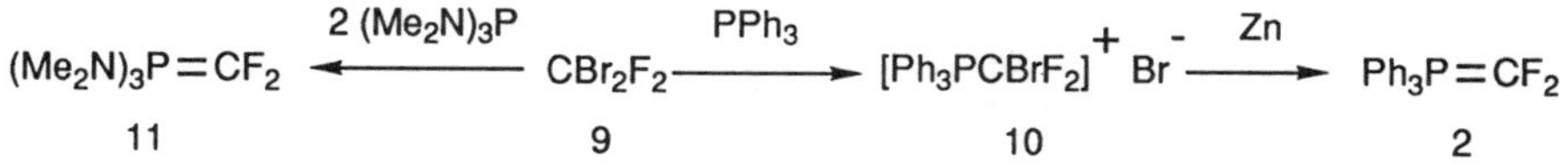

Scheme 3

treatment with $(Me_3N)_3P/CF_2Br_2/Zn$ (Scheme 6) *(17)*. The silylated *(18)* or acylated *(19)* forms of 1,1-difluoro-1-alken-3-ols (19 and 20) were obtained from the respective silyl or acyl protected α-hydroxy -aldehyde or -ketone *(19)* by reaction with the Wittig reagent generated *in situ* with $CF_2Br_2/(Me_2N)_3P$ (Scheme 7).

Perfluoroalkyl acyl fluorides react with the fluoro diphosphonium salt 22 to provide vinyl phosphonium salts (23) in good yields *(20)* (Scheme 8); hydrolysis gave (*E*)-1,2-difluoro-olefins (24). Alternatively, 22 could be reacted with aldehydes to provide the fluoroalkenylphosphonium salt (25) which, when treated with aqueous sodium hydroxide, affords a convenient one pot synthesis of fluoroolefins (26)*(21)*. The unexpected Z stereoselectivity observed in the synthesis of 26a (*E/Z* = 13/87) was rationalized on the basis of through space charge-transfer complexes between the pi electrons of the aromatic ring of the aldehyde and the positive charge of one of the tri-n-butylphosphonium groups.

The Wadsworth-Emmons reagent 27 has been used successfully to prepare difluoroolefins (3) (Scheme 9) *(22,23)*. Furthermore, when the trimethylsilyl derivative 30 is used in place of 27, the condensation reaction can be performed under very mild conditions. In these reactions, the intermediate 29 does not collapse spontaneously, but requires heating to liberate 3. Edwards et *al.* *(24)* have found that 33, a stable, readily prepared crystalline compound is a convenient alternative to 27 and 30 (Scheme 10). It is interesting to note that they were unable to utilize the monofluoro reagent 36 to obtain monofluoroolefins, presumably because of the decomposition of 36 *via* a carbene intermediate. However, Blackburn *(25)* reported that monofluoro reagent 34 was useful for achieving this transformation, but no yields were given (Scheme 11). Fujita and Hiyama *(26)* have shown that the thermally stable zinc reagent, CF_3CCl_2ZnCl, will add to aldehydes (37) affording difluoroolefins (39) in good yields (Scheme 12).

Recently *(27)*, the Wadsworth-Emmons reagent (41), generated *in situ* from 40, was shown to react with aldehydes and ketones providing the conjugated fluoro vinyl sulfones 42 in good to excellent yields (Scheme 13). Amalgamated alumina reductively removes the phenylsulfonyl group to afford an equimolar mixture of the E and Z fluoroolefins 3 *(28)*. The anion generated from α-fluorosulfone 40 condenses directly with aromatic aldehydes to give 44 which can be elaborated to β-fluorostyrenes (46) (Scheme 14). However, this reaction appears to be limited to aromatic aldehydes.

1,1-Difluoroethylene (47) is a useful building block for the preparation of 1,1-difluoroolefins. The vinylic fluorine atoms stabilize the difluorovinyl lithium 48 formed by treatment of 47 with *sec*-BuLi at -115°C. This carbanion can be quenched with aldehydes *(19,29,30)* or carbon dioxide *(31)* to provide 1,1-difluoroolefins (49,50) which are of utility for further transformations (Scheme 15). Similarly, trifluoroethylene (51) has been utilized to prepare 1,2-difluoroolefins, such as 53 (Scheme 16) *(32)*. Difluoroacetylene 52 was implicated as an intermediate. Kolb et *al.* *(33)* prepared the trifluoroolefin 55 from aldehyde 54 on route to the amino acid 56 (Scheme 17).

12 → 13

Scheme 4

14 → 15 (CF_2Br_2, $(Me_2N)_3P$, Triglyme, 85°, 95%)

Scheme 5

16 → 17 (CF_2Br_2, $(Me_3N)P$)

Scheme 6

18 → 20 (CF_2Br_2 / $P(NMe_2)_3$; X = Ac, R' = alkyl)

18 → 19 (CF_2Br_2 / $P(NMe_2)_3$; X = $SiMe_3$, R' = H)

Scheme 7

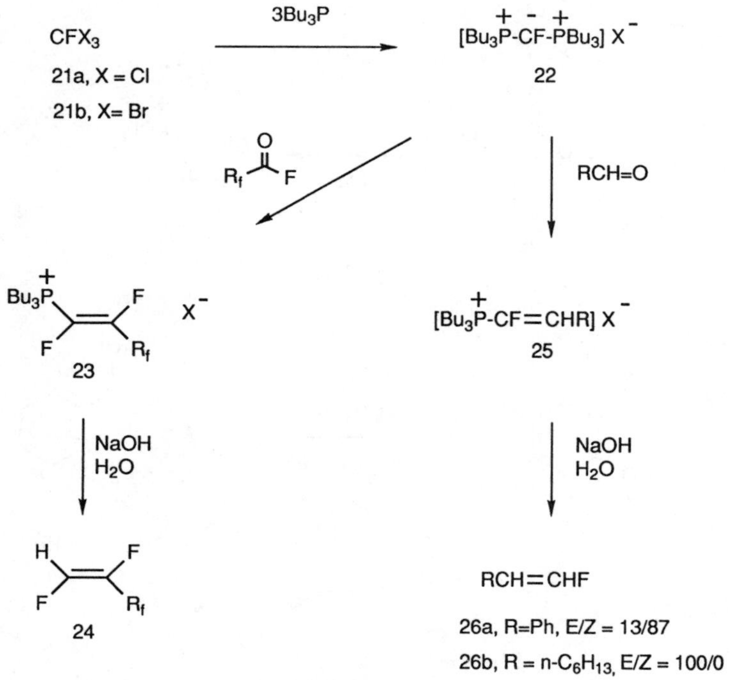

CFX_3 —($3Bu_3P$)→ $[Bu_3\overset{+}{P}\text{-}\overset{-}{C}F\text{-}\overset{+}{P}Bu_3]\,X^-$

21a, X = Cl
21b, X= Br

22

22 —($R_fC(O)F$)→ 23: $Bu_3\overset{+}{P}(F)C{=}C(F)R_f\ X^-$

22 —(RCH=O)→ 25: $[Bu_3\overset{+}{P}\text{-}CF{=}CHR]\,X^-$

23 —(NaOH, H_2O)→ 24: $H(F)C{=}C(F)R_f$

25 —(NaOH, H_2O)→ RCH=CHF

26a, R=Ph, E/Z = 13/87
26b, R = n-C_6H_{13}, E/Z = 100/0

Scheme 8

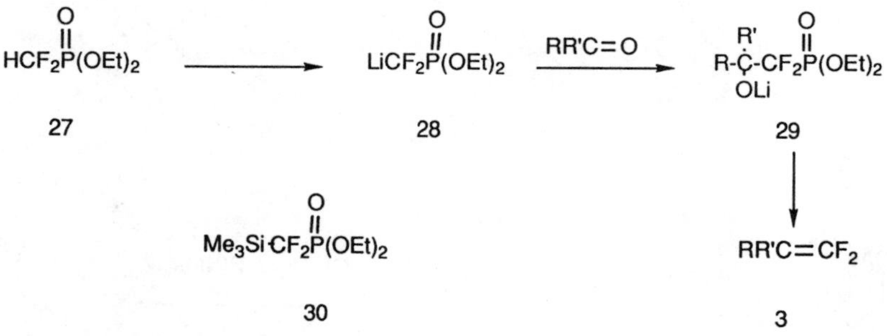

$HCF_2P(O)(OEt)_2$ (27) → $LiCF_2P(O)(OEt)_2$ (28) —(RR'C=O)→ $R\text{-}C(R')(OLi)\text{-}CF_2P(O)(OEt)_2$ (29) → $RR'C{=}CF_2$ (3)

$Me_3Si\text{-}CF_2P(O)(OEt)_2$

30

Scheme 9

$HP(O)Ph_2$ (32) → $Ph_2P(O)CF_2H$ (33) → → $RR'C{=}CF_2$ (3)

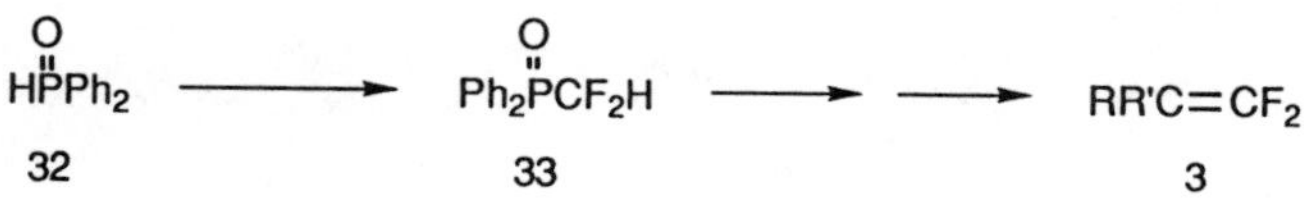

Scheme 10

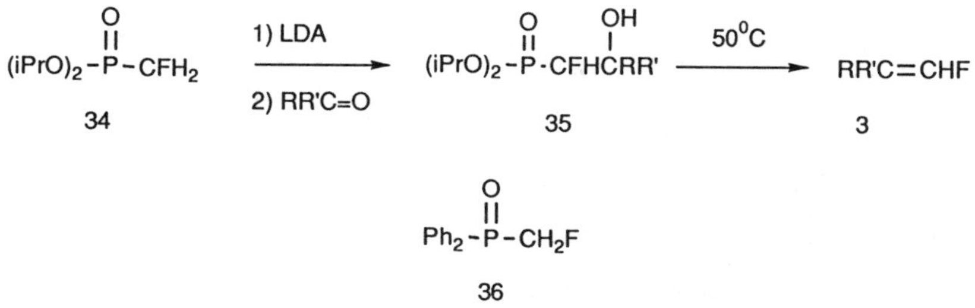

$$(iPrO)_2-P(O)-CFH_2 \xrightarrow[\text{2) RR'C=O}]{\text{1) LDA}} (iPrO)_2-P(O)-CFHC(OH)RR' \xrightarrow{50^0C} RR'C{=}CHF$$

34 → 35 → 3

$$Ph_2-P(O)-CH_2F$$

36

Scheme11

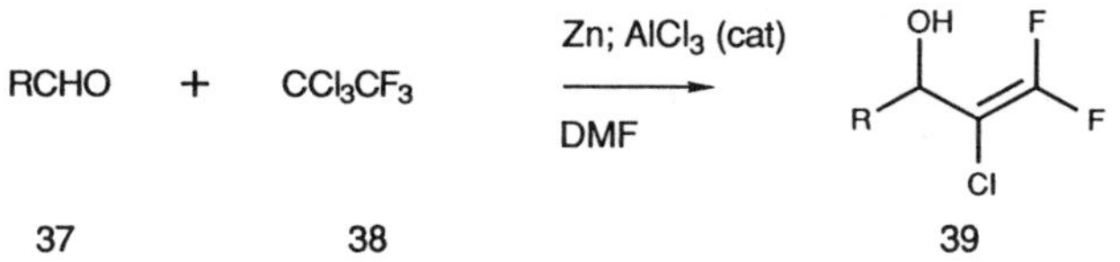

$$RCHO + CCl_3CF_3 \xrightarrow[\text{DMF}]{\text{Zn; AlCl}_3\text{ (cat)}} RCH(OH)C(Cl){=}CF_2$$

37, 38, 39

Scheme 12

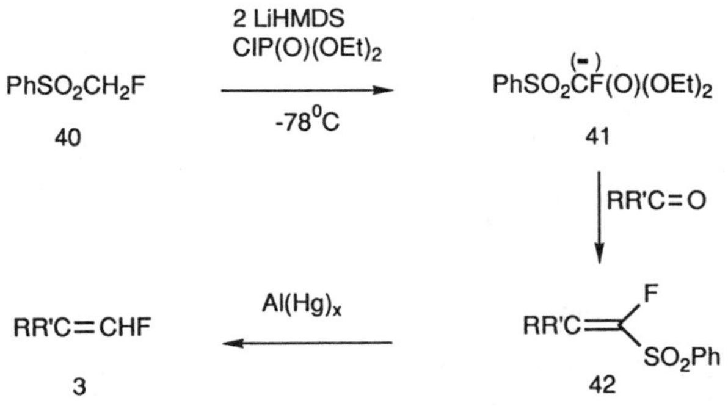

$$PhSO_2CH_2F \xrightarrow[-78^0C]{\text{2 LiHMDS, ClP(O)(OEt)}_2} PhSO_2\overset{(-)}{C}F(O)(OEt)_2$$

40 → 41

$$\xrightarrow{RR'C=O} RR'C{=}CF(SO_2Ph) \xrightarrow{Al(Hg)_x} RR'C{=}CHF$$

42 → 3

Scheme 13

$$PhSO_2CH_2F \xrightarrow{BuLi} PhSO_2CHLiF \xrightarrow{ArCH=O} ArCH(OH)CHF(SO_2Ph)$$

40 → 43 → 44

$$\xrightarrow{MsCl,\ NEt_3} Ar(H)C{=}CF(SO_2Ph) \xrightarrow{Al(Hg)_x} Ar(H)C{=}CHF$$

45 → 46

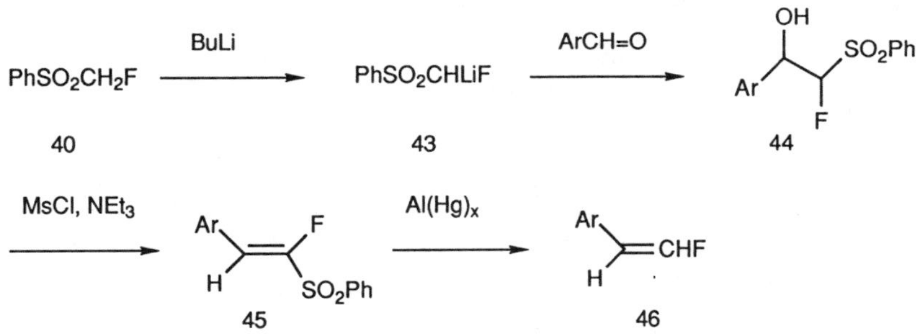

Scheme 14

Burton *(34)* recently reported that iodotrifluoroethylene (58) couples directly with 1-alkynes (57) in the presence of palladium, cuprous iodide and triethylamine to give excellent yields of trifluoroenynes (59) (Scheme 18). The difluoroenolphosphonate 61, prepared from chlorodifluoromethylaryl ketones (60), can be metalated with Bu_2CuLi in the presence of TMEDA and the resulting carbanion trapped by allyl halides to afford difluorostyrenes (62) (Scheme 19) *(35)*.

The direct introduction of a fluorine atom is a rare reaction in the context of preparing fluoroolefins. Lee and Schwartz *(36)*, however, have developed a stereospecific method to fluoroolefins from the corresponding vinyl iodide by metal-halogen exchange at low temperature followed by the addition of an *N*-fluoro-*N*-alkylsulfonamide (Scheme 20). The temperature and solvent are critical to the success of the reaction. The olefin (66) was obtained as a by-product.

Halofluorohydrocarbons as Precursors to Fluoroolefins

A general method leading to the stereospecific synthesis of 3-fluoroacrylate derivatives arose from the early work of Shen et *al.* *(37)* who demonstrated that chlorodifluoromethane adds to malonate carbanions to give α-difluoromethylmalonate derivatives (Scheme 21). Bey and co-workers *(38-40)* made extensive use of this reaction in a major program directed towards the synthesis of α-difluoromethylamines and amino acids. A key step in these syntheses is the acid catalyzed decarboxylation of the α-difluoromethylmalonates (70) (Scheme 22). On occasion, 3-fluoroacrylates (72) were observed as side-products in addition to the expected compounds (71).

If the decarboxylation step of the malonate hemiesters is performed in the presence of one equivalent of base, clean decarboxylative-halide elimination occurs leading to high yields of 72 *(41)* (Schemes 22, 23). Furthermore, the decarboxylative-halide elimination reaction proceeded stereospecifically to yield exclusively the *(E)* isomer (75) of the fluoroolefin. While the underlying mechanism behind this stereocontrol is uncertain, some years earlier Krapcho *(42)* had reported that decarboxylative-bromide and -chloride eliminations resulted upon saponification of diethyl dibromo(and dichloro)malonates to yield a single isomer of the haloolefin. In similar experiments aimed at the synthesis of the prostaglandin derivative 77, Kosuge et *al.* *(43)* confirmed the stereochemical outcome of the decarboxylative-fluoride elimination reaction (Scheme 24). In a more complicated system (Scheme 25), however, a mixture of isomers (79,80) was obtained; the harsh reaction conditions may be the reason for this apparent loss of stereocontrol. Tsushima and Kawada (45) used a variation of this reactions to prepare 3-fluoropyruvic acid (84) (Scheme 26). The decarboxylative-fluoride elimination reaction has been extended to the preparation of 3,3-difluoroacrylates (87) through the reaction of dibromodifluoromethane with malonates under basic conditions (Scheme 27) *(46)*. Mono- and di-fluoroacrylates were easily elaborated to the corresponding fluoroallylamines (41).

While the decarboxylative-halide elimination procedure provides a general approach to β-fluoroacrylates, problems arise in the initial addition reaction where the malonate derived anion (88) is substituted with a leaving group in the β-position (88).

sec-BuLi, -115^{0}; RCH=O; CO_2

47 48 49 50

Scheme 15

t-BuLi; [FC≡CF]; CO_2Me; LDA, HMPA

51 52 53

Scheme 16

CO_2tBu; OH; NH_2; CO_2H

54 55 56

Scheme 17

RC≡CH + ; $Pd(PPh_3)_2Cl_2$; CuI, Et_3N

57 58 59

Scheme 18

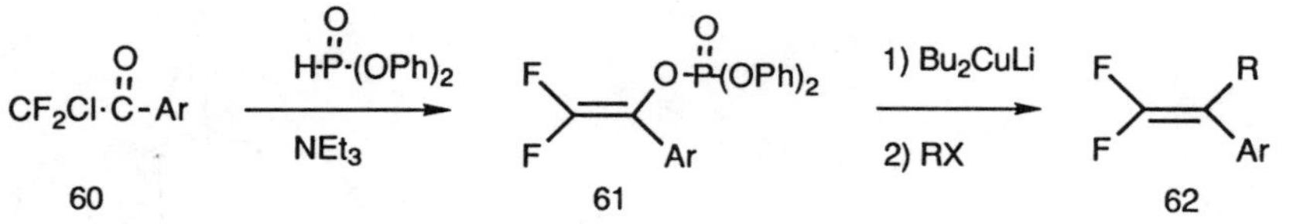

$CF_2Cl \cdot C(O) - Ar$ (60) $\xrightarrow[NEt_3]{HP(O)(OPh)_2}$ $F_2C{=}C(Ar)OP(O)(OPh)_2$ (61) $\xrightarrow[2)\ RX]{1)\ Bu_2CuLi}$ $F_2C{=}C(R)Ar$ (62)

Scheme 19

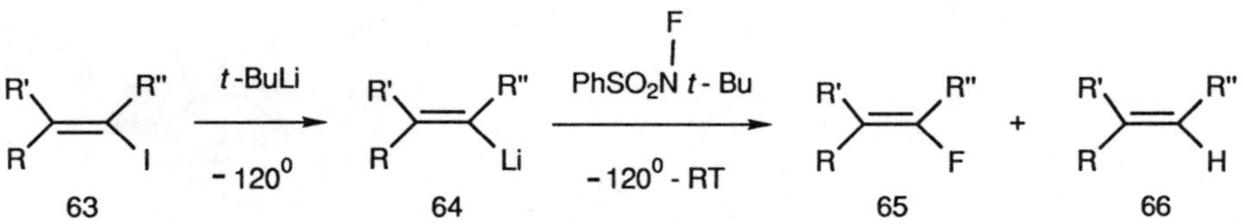

$RR'C{=}C(R'')I$ (63) $\xrightarrow[-120^0]{t\text{-BuLi}}$ $RR'C{=}C(R'')Li$ (64) $\xrightarrow[-120^0\ -\ RT]{PhSO_2N(F)\,t\text{-Bu}}$ $RR'C{=}C(R'')F$ (65) + $RR'C{=}C(R'')H$ (66)

Scheme 20

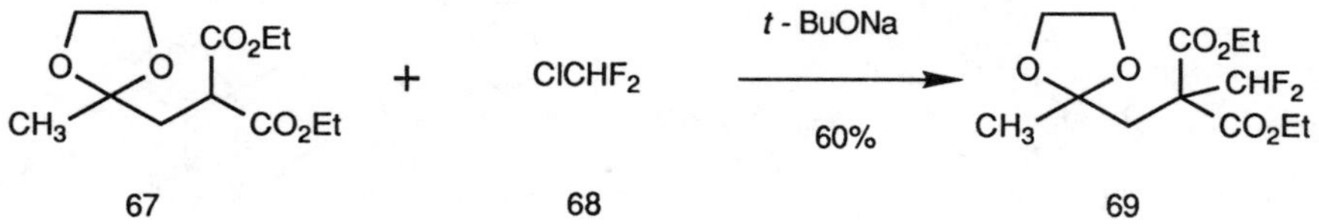

67 + $ClCHF_2$ (68) $\xrightarrow[60\%]{t\text{-BuONa}}$ 69

Scheme 21

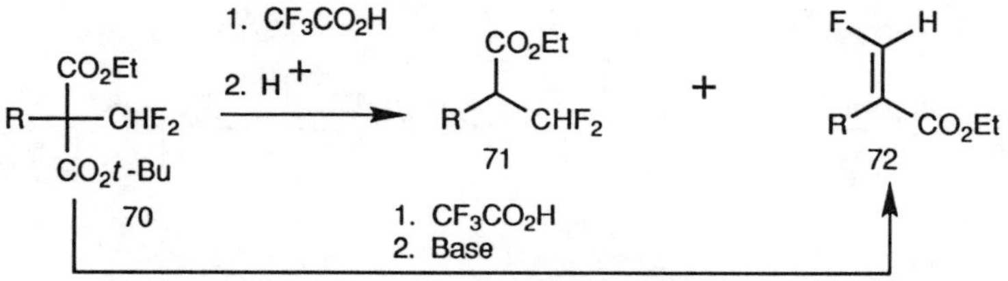

$R-C(CO_2Et)(CO_2t\text{-Bu})-CHF_2$ (70) $\xrightarrow[2.\ H^+]{1.\ CF_3CO_2H}$ $R-CH(CO_2Et)-CHF_2$ (71) + $FHC{=}C(R)CO_2Et$ (72)

70 $\xrightarrow[2.\ Base]{1.\ CF_3CO_2H}$ 72

Scheme 22

$R-X-CH_2-C^-(CO_2Et)-CO_2t\text{-Bu}$

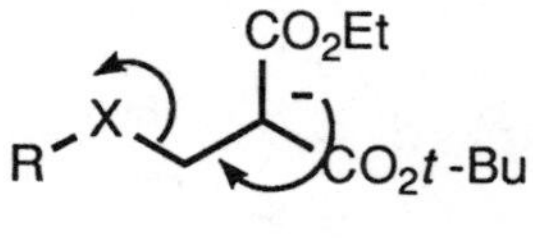

88

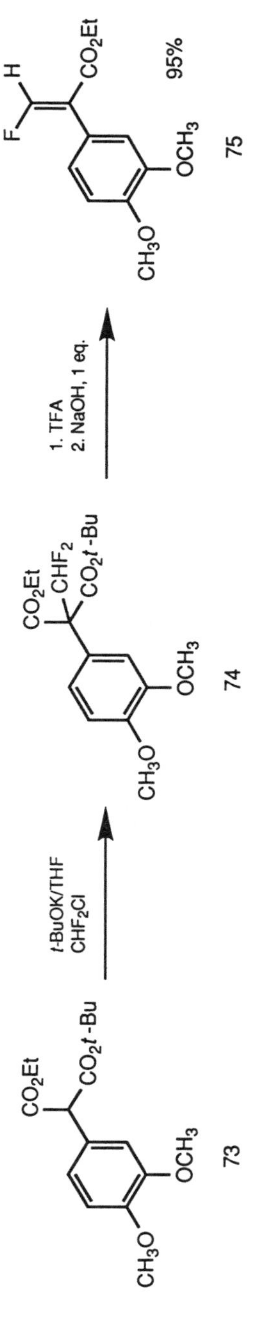

73: CH_3O, OCH_3, CO_2Et, CO_2t-Bu

t-BuOK/THF, CHF_2Cl →

74: CH_3O, OCH_3, CO_2Et, CHF_2, CO_2t-Bu

1. TFA
2. NaOH, 1 eq. →

75: CH_3O, OCH_3, F, H, CO_2Et — 95%

Scheme 23

76: CO_2Et, CHF_2, CO_2Et

1. KOH, EtOH/DMF
2. KF/DMF, heat →

77: F, H, CO_2Et

Scheme 24

78: HCONH, N, S, CO_2Me, CHF_2, CO_2Me

LiI, DMF, 160°C →

79: HCONH, N, S, F, H, CO_2Me — 52%

\+

80: HCONH, N, S, H, F, CO_2Me — 14%

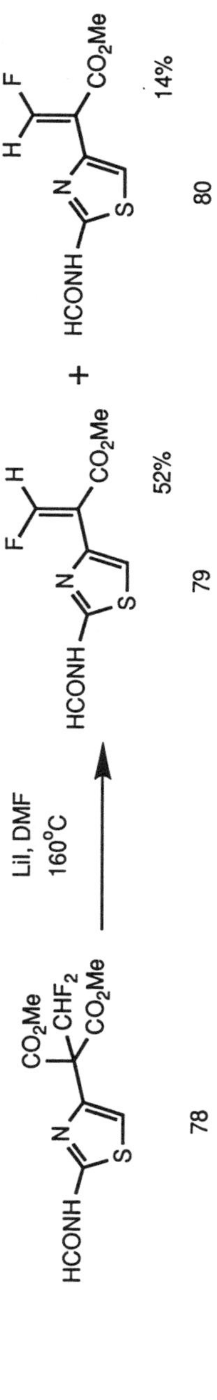

Scheme 25

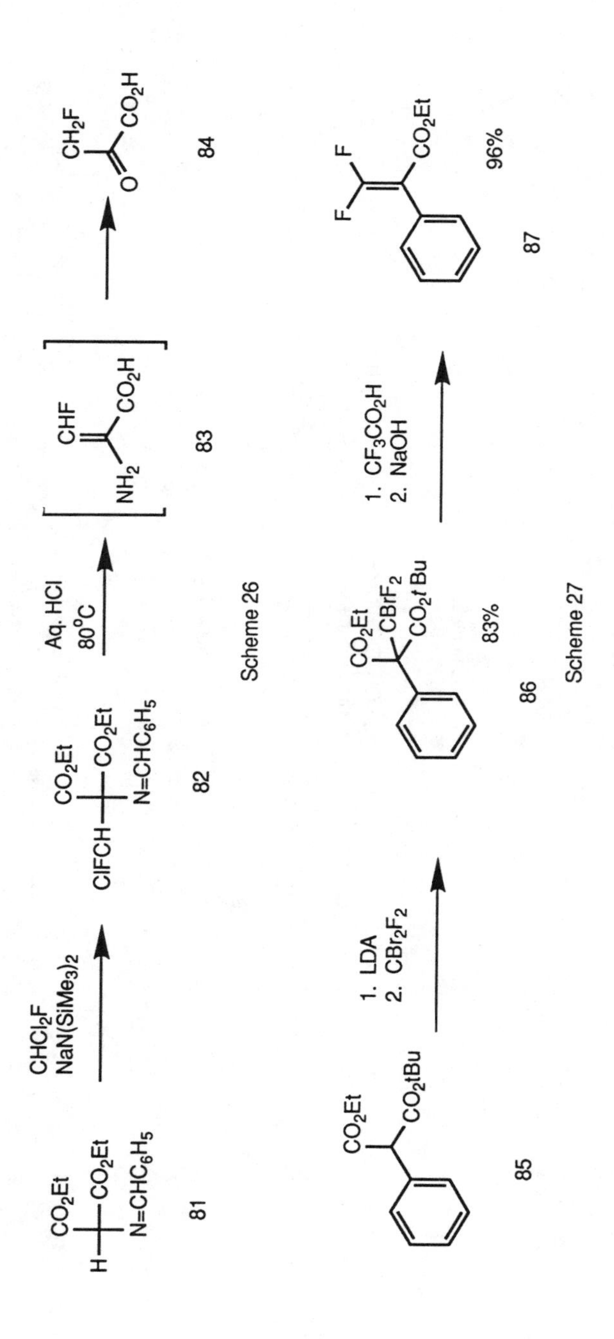

Scheme 26

Scheme 27

This shortcoming was overcome by preparing the isomeric synthons **92** and **94** which can be reacted with a number of nucleophiles *(47)* (Scheme 28). The key reaction was the allylic bromination of the (*E*)-vinyl fluoride (**91**) available in eight steps from the malonate **89**. Interestingly, isomerization of the vinyl fluoride was observed during the allylic bromination reaction.

Since not even traces of the (Z)-fluoroacrylates were observed in the decarboxylative-fluoride elimination step of the α-difluoromethylmalonate derivatives, this isomer was prepared by isomerization of the (*E*)-vinyl fluorides. For example, the phthalimido derivative **96** was isomerized in a straightforward two-step procedure to afford a separable mixture of the (*E*) (**96**) and (Z) (**98**) isomers *(41)* (Scheme 29).

In another approach to 3,3-difluoroacrylate derivatives, Wakselmann and his group *(48,49)* have taken advantage of the observation that dibromodifluoromethane adds to ynamines (**99**) (Scheme 30) and enol ethers (**102**) (Scheme 31) under UV irradiation. Subsequent elaboration of the initial addition products, **100** and **103**, derived from *N*,*N*-diethyl-1-butynylamine and ethyl vinyl ether afforded 3,3-difluoroacrylamide (**101**) and 3,3-difluoroacrylate (**106**), respectively.

Elimination and Isomerization Approaches to Fluoroolefins

Elimination, reductive elimination, reductive alkylation, and isomerization reactions have been employed in order to create the fluoroolefin moiety in molecules already containing one or more fluorine atoms. In general, these reactions were developed for a specific purpose and are often not generally applicable to other systems. For example, base catalyzed HF elimination converted 1,2,2-trifluoroethylcyclohexane (**107**) to the difluoroolefin (**108**)*(50)* (Scheme 32). Reductive elimination reactions were employed to prepare the interesting fluoroallene (**110**) *(51)* (Scheme 33). Similarly, **111** affords a mixture of **112** and **113** *(52)* (Scheme 34).

An alternate route to fluoroolefins relies upon the ease of reduction of difluoroolefins(*18*). Reduction of **114** with sodium bis(2-methoxyethoxy)aluminum hydride (Scheme 35) afforded the fluoroolefins **115** and **116** considerably enriched with the (*E*)-isomer **116**. In a complementary reaction, reduction of the allylic alcohol **117** with $LiAlH_4$ afforded selectively the (Z)-isomer **118**. The difluoromethacrylic acid (**121**) was prepared in similar manner from **120** (Scheme 36) *(53*; for related examples see references *75* and *76)*. Under more forcing conditions, further reduction afforded 3-fluoromethacrylic acid **122**. Of more general use is the reaction of **120** with Grignard reagents whereupon the 1,4-addition elimination mechanism offers an entry into α-difluoromethylene substituted aliphatic and aromatic carboxylic acids **123**. Ester enolates (**125**) have been shown to add to trifluoropropene (**124**) forming the difluoroolefins (**126**) (Scheme 37) *(54)*.

Ichikawa et al. *(55)* have reported a general procedure for the synthesis of hydrocarbons substituted with a terminal difluoroolefin (Scheme 38). Elimination of HF from the tosylate **127** afforded a lithium species **128** which was elaborated to **129** using organoborane chemistry. The yields were generally good but the nature of the reaction precludes many functional groups.

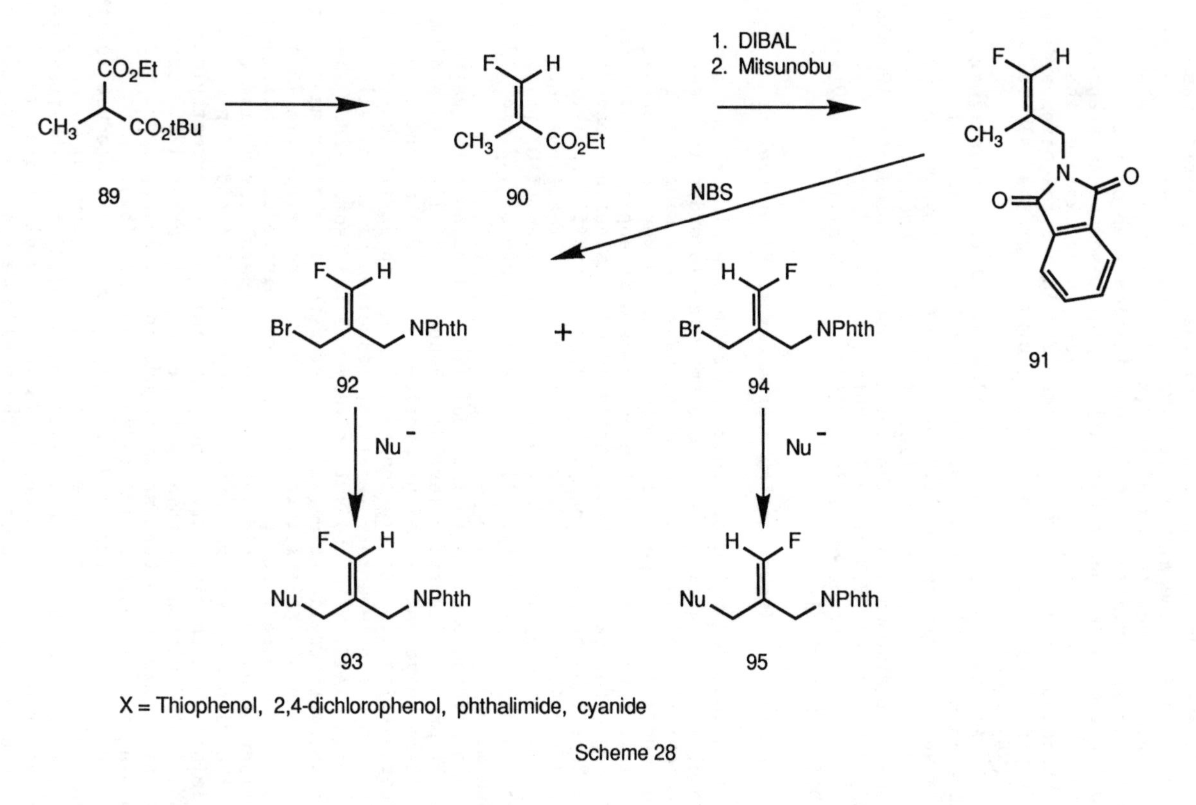

X = Thiophenol, 2,4-dichlorophenol, phthalimide, cyanide

Scheme 28

Scheme 29

Scheme 30

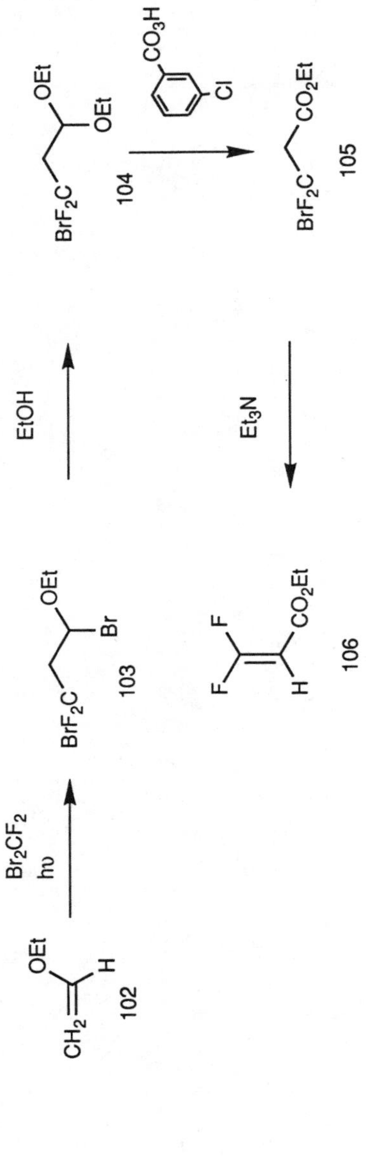

$CH_2{=}CH(OEt)$ **102** $\xrightarrow{Br_2CF_2,\ h\nu}$ $BrF_2CCH_2CH(Br)OEt$ **103** $\xrightarrow{EtOH}$ $BrF_2CCH_2CH(OEt)_2$ **104** $\xrightarrow{m\text{-}ClC_6H_4CO_3H}$ $BrF_2CCH_2CO_2Et$ **105** $\xrightarrow{Et_3N}$ $F_2C{=}CHCO_2Et$ **106**

Scheme 31

$C_6H_{11}CF_2CH_2F$ **107** $\xrightarrow{t\text{-BuOK}}$ $C_6H_{11}CF{=}CHF$ **108**

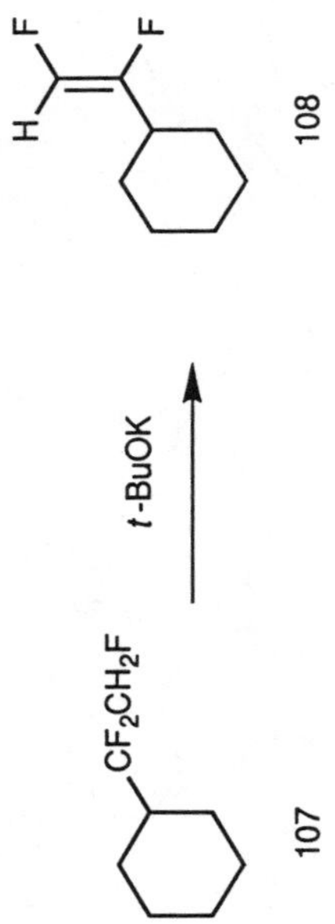

Scheme 32

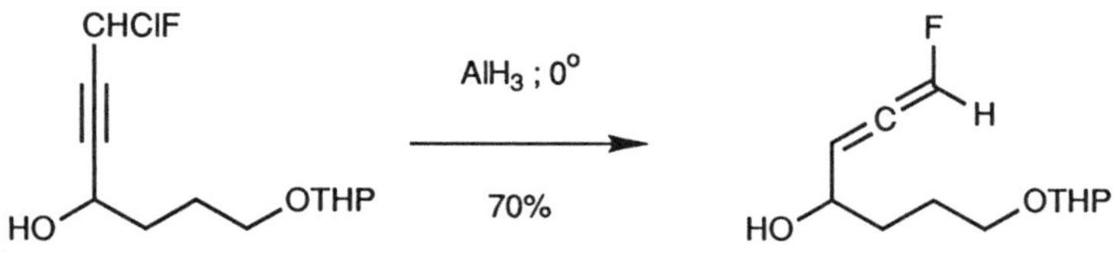

CHClF

AlH_3 ; 0°

70%

HO OTHP

F H HO OTHP

109 110

Scheme 33

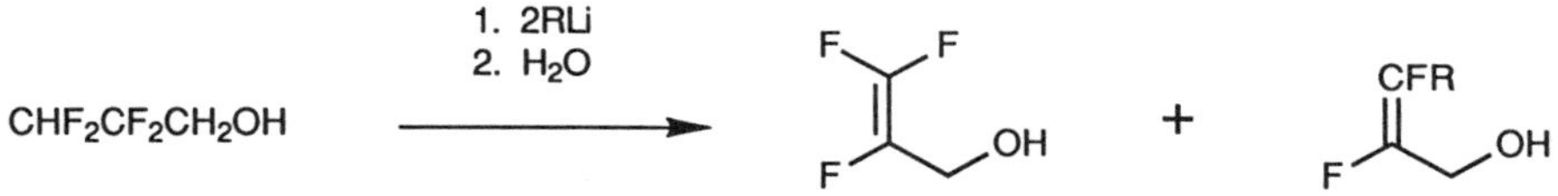

1. 2RLi
2. H_2O

$CHF_2CF_2CH_2OH$

F F F OH + CFR F OH

111 112 113

Scheme 34

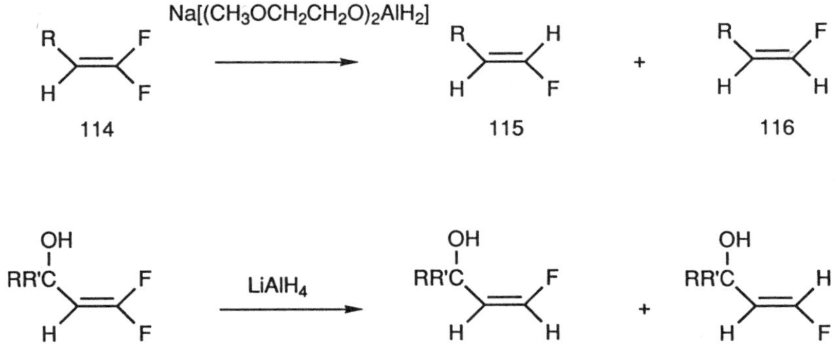

$Na[(CH_3OCH_2CH_2O)_2AlH_2]$

R F H F → R H H F + R F H H

114 115 116

OH RR'C F H F → LiAlH₄

$LiAlH_4$

OH RR'C F H H + OH RR'C H H F

117 118 119

Scheme 35

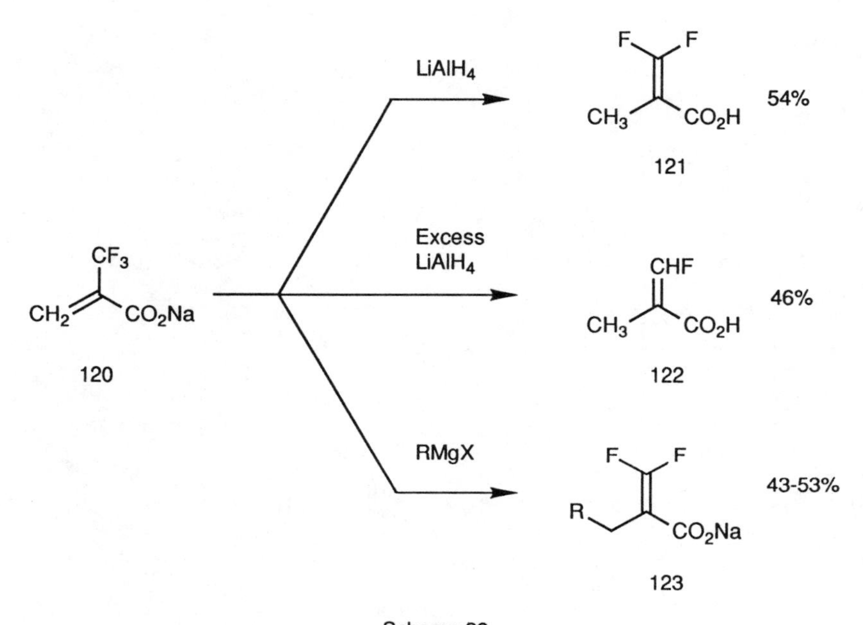

Scheme 36

CF_3 (124) + $R^1R^2CHCO_2R^3$ (125) → 1. LDA, -78°C; 2. Room temp. → 126 (R^1, R^2, CO_2R^3, F, F)

124 125 126

Scheme 37

CF_3CH_2OTs (127) → 2 eq. LDA → 128 (F, F, TsO, Li) → 1. BR_3; 2. H^+ → 129 (F, F, R, H)

127 128 129

Scheme 38

Difluoroenolates have also been used as intermediates to α,α-difluorocarbonyl compounds *via* a Claisen rearrangement. Metcalf et *al.* *(56)* (Scheme 39) prepared difluorovinyl ethers (131) by base treatment of 130 and found that spontaneous rearrangement to the difluoroaldehyde (132) occurred under the reaction conditions. In similar fashion (Scheme 40) 133 was converted to 135 *via* the enol 134 *(30)*. Alternatively, the enolate can also be trapped as the trimethylsilyl ether (137) prior to rearrangement (Scheme 41) *(59)*.

In an interesting transformation, Nakai et *al.* *(58)* treated the sulfoxide (140) with base to generate the fluorovinylsulfoxide (141), then showed that allylic alcohols (139) add, *via* an addition-elimination sequence, and the resulting enol (142) rearranges spontaneously to 143 (Scheme 42).

α-α-Difluoro-O-silyl ester enolates have also found utility in other condensation reactions. For example, the ester enolate (144) condenses with aldehydes (145) to form α,α-difluoro-β-hydroxy esters (146) (Scheme 43) *(59)* or with imines (147) to generate difluoroazetidin-2-ones (148) (62). Lewis acid catalyzed condensation readily occurs between the enol (149) and ketones or aldehydes (Scheme 44)*(61)* affording α,α-difluoro-β-hydroxy ketones (151).

Deconjugation of 4-fluorocrotonates (152) has afforded an entry into the fluoroolefins 153 which were used for the synthesis of number of β-fluoromethylene α-amino acids (154) *(62,63)*(Scheme 45). Predominantly a single isomer ((E) > 99%) is obtained, although the overall conversion of 152 to 154 is quite low. These amino acids are also obtainable from the diastereomeric mixture of oxazolones 155 (Scheme 46) *via* a reductive elimination procedure. Whereas the yields were high in this case, there was very little control of stereochemistry *(62)*.

The Fluoro-Pummerer and Related Reactions

Although general synthetic methods to disubstituted fluoroolefins had been developed by the early 1980's, monosubstituted analogs were not readily accessible until Reutrakul et *al.*(*64*) demonstrated that pyrolysis of α-fluorosulfoxides (161) provided fluoroolefins in good yield, albeit as a mixture of isomers (Scheme 47). The utility of this procedure was further enhanced by the report *(65,66)* of a facile synthesis of the key intermediate 159. Treatment of methyl phenyl sulfoxide with diethylaminosulfur trifluoride (DAST), in the presence of antimony trichloride provides 159 in quantitative yield *(66)*. The reaction proceeds in good yield with dialkyl sulfoxides and alkyl aryl sulfoxides (163). Reoxidation of the α-fluorosulfide (165) to the corresponding sulfoxide (161), followed by pyrolysis, provides a direct synthesis of fluoroolefins *(65)*. The reaction is believed to proceed by a Pummerer-type mechanism (*i.e.*, a fluoro-Pummerer reaction, Scheme 48). Similarly, Umemoto *(67)* reported that *N*-fluorocollidine (167) converted sulfides to α-fluorosulfides (170) presumably via an *S*-fluorosulfonium cation species 168 (Scheme 49). The synthetically challenging fluorovinyl ether nucleosides (175) and (176) were prepared using the fluoro-Pummerer reaction (Scheme 50) *(68)*; the (*E*)-isomer (175) could be isomerized to 176 under photolytic conditions. Finch and co-workers *(69)* converted 160 to the sulfoximine 178 and demonstrated the utility of this compound as a mild fluoromethylene synthon (Scheme 51). Base-catalyzed condensation 178 with a carbonyl compound gave 179 which afforded

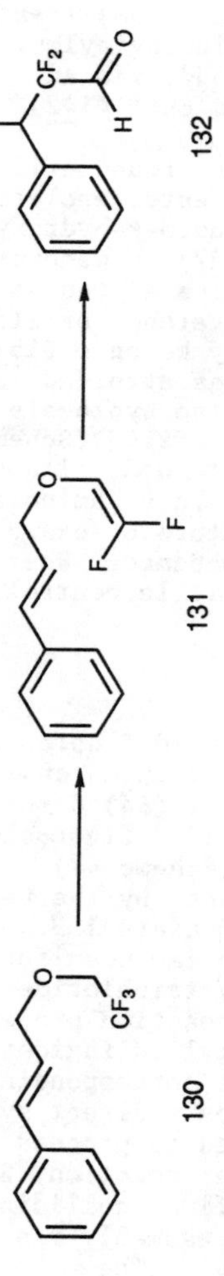

130 → 131 → 132

Scheme 39

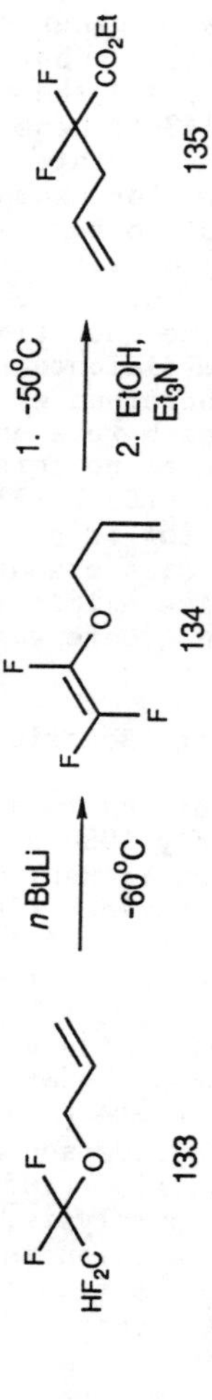

133 —(nBuLi, -60°C)→ 134 —(1. -50°C; 2. EtOH, Et_3N)→ 135

Scheme 40

136 —(Zn, $ClSiMe_3$)→ 137 —(1. H_2O/ SiO_2; 2. EtOH)→ 135

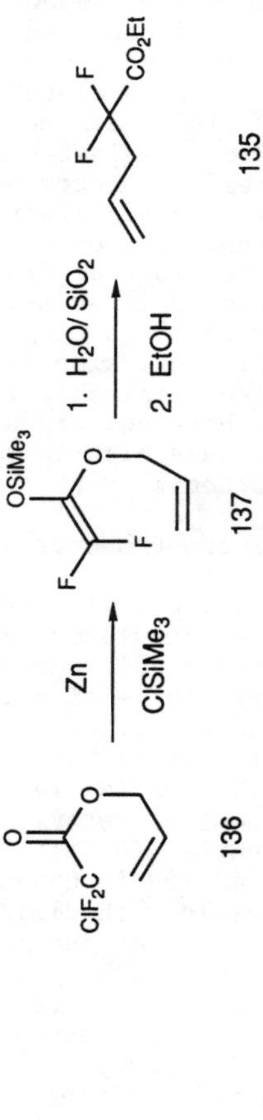

Scheme 41

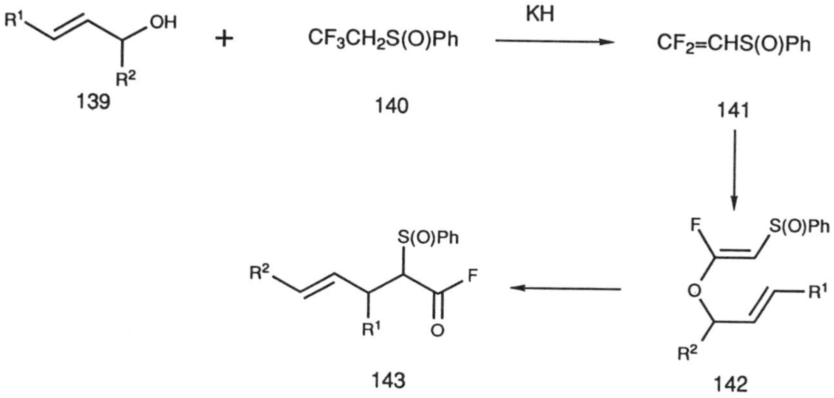
R1
OH
R2
139
+
CF3CH2S(O)Ph
140
KH
CF2=CHS(O)Ph
141
F
S(O)Ph
O
R1
R2
142
S(O)Ph
R2
F
R1
O
143

Scheme 42

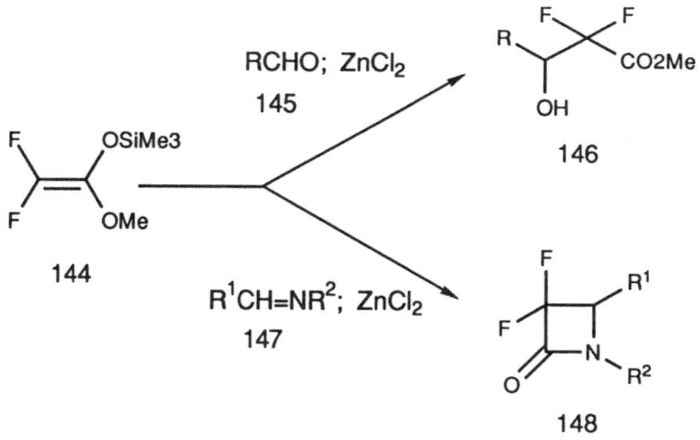
F
OSiMe3
F
OMe
144
RCHO; ZnCl2
145
F
F
R
CO2Me
OH
146
R1CH=NR2; ZnCl2
147
F
R1
F
N
R2
O
148

Scheme 43

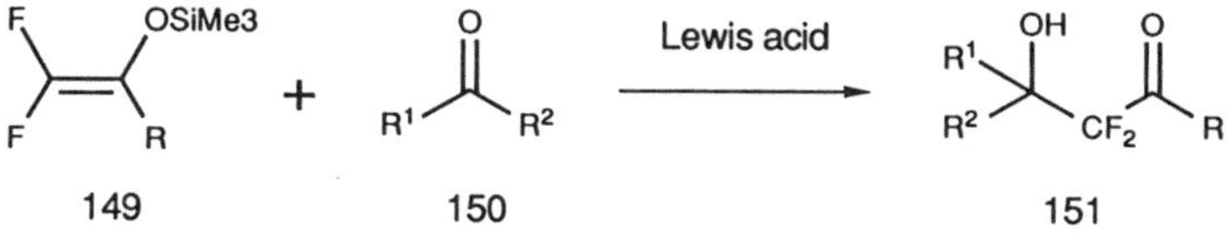
F
OSiMe3
F
R
149
+
O
R1
R2
150
Lewis acid
OH
O
R1
R2
CF2
R
151

Scheme 44

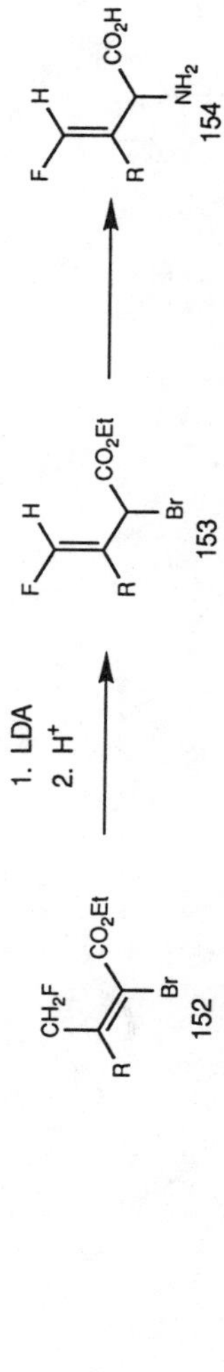

CH_2F CO_2Et R Br 152 — 1. LDA 2. H^+ → F H R CO_2Et Br 153 → F H R CO_2H NH_2 154

R = CH_3, or substituted aromatic groups

Scheme 45

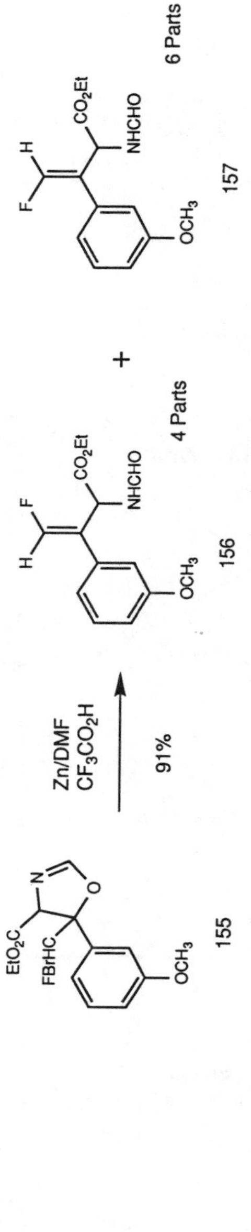

EtO_2C FBrHC N O OCH_3 155 — Zn/DMF CF_3CO_2H, 91% → H F CO_2Et NHCHO OCH_3 156 (4 Parts) + F H CO_2Et NHCHO OCH_3 157 (6 Parts)

Scheme 46

$ClCH_2SPh$ (158) — CsF, CH_3CN, 18-Crown-6, 100 hr → FCH_2SPh (159) — NBS, H_2O → $FCH_2S(O)Ph$ (160) — 1) LDA; -78°C 2) RCH_2X → $RCH_2CHFS(O)Ph$ (161) — 180°C → RCH=CHF (162)

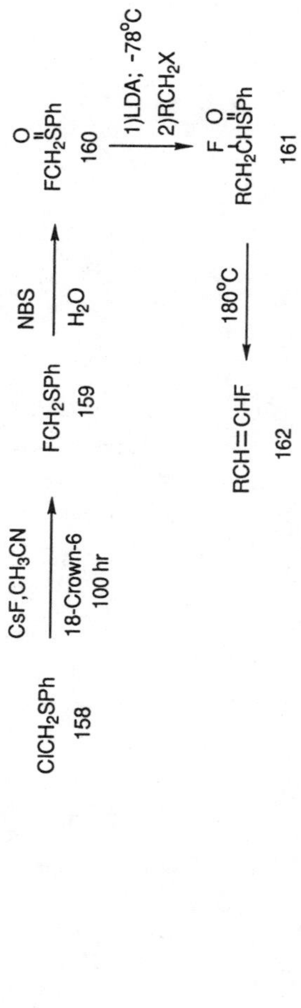

Scheme 47

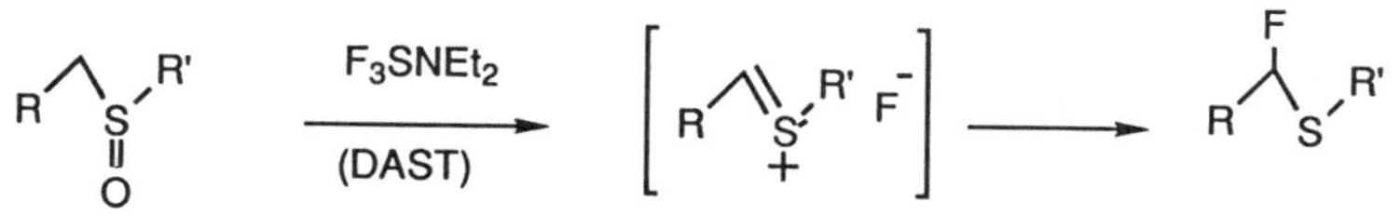

$R\text{-}CH_2\text{-}S(=O)\text{-}R'$ (163) $\xrightarrow[\text{(DAST)}]{F_3SNEt_2}$ [$R\text{-}CH=S^+\text{-}R'$ F^-] (164) ⟶ $R\text{-}CHF\text{-}S\text{-}R'$ (165)

Scheme 48

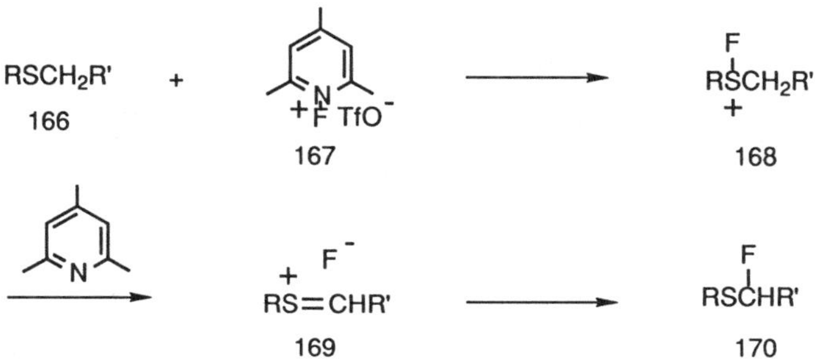

$RSCH_2R'$ (166) + 2,4,6-trimethyl-N-fluoropyridinium TfO^- (167) ⟶ $RS^+(F)CH_2R'$ (168) $\xrightarrow{\text{2,4,6-collidine}}$ $RS^+{=}CHR'$ F^- (169) ⟶ $RSCHFR'$ (170)

Scheme 49

171 (ArS(O), Ad) $\xrightarrow{\text{DAST}}$ 172 (ArS, F, Ad) $\xrightarrow{\text{MCPBA}}$ 173 (ArS(O), F, Ad) $\xrightarrow[\text{Hunig's base}]{140^0, \text{ diglyme}}$ 174 (HFC=, Ad) $\xrightarrow[\text{Dowex column}]{TFA/H_2O}$ 175 (H, F; Ad; HO, OH) + 176 (F, H; Ad; HO, OH)

176 $\xleftarrow{h\nu, \text{ acetone}}$ 175

Scheme 50

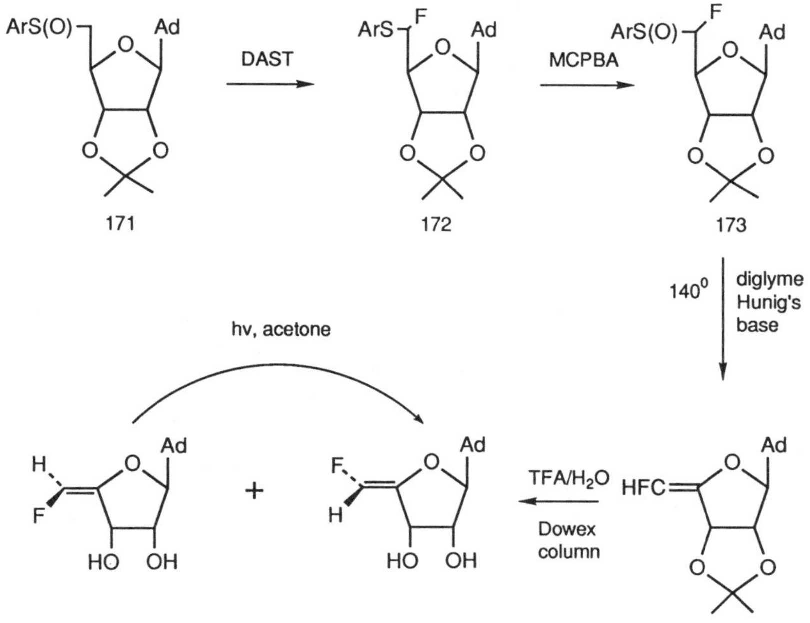

fluoroolefins (180) upon a reductive elimination. α-Fluorosulfides (182), formed by treatment of thioacetates with HgF_2 in CH_3CN, can be oxidized to sulfoxides and pyrolized to give fluoroolefins (183) (70) (Scheme 52).

Fluoroolefin-Containing Enzyme Inhibitors

After the initial discovery that β-fluoromethylene-substituted amines (e.g., 184, Table 1) were potent, mechanism-based inhibitors of monoamine oxidase (MAO) (*41*), the concept was successfully broadened to include most of the common amine oxidases (Table 1). This approach was also used to design inhibitors of γ-aminobutyric acid transaminase; both the α- and β- substituted amino acids 189 and 190 were found to inactivate this enzyme. Recently, application of this concept to the design of inhibitors of S-adenosylhomocysteine hydrolase (SAH) has led to the discovery of very potent inhibitors of this enzyme (e.g., 176, Table 1).

The fluorine atom, when compared to chlorine or hydrogen atoms, confers unique potency to the enzyme inhibitor as exemplified by the inactivation of MAO and SAH hydrolase (Table 2). It is interesting to note that whereas MAO catalyzes the oxidative transformation of an amine to an aldehyde, SAH hydrolase proceeds with overall net retention of oxidation state. However, advantage has been taken of a critical oxidation step along the mechanistic pathway of this enzyme to design inhibitors such as 176. β-Methylenephenethylamine (190, Table 2) is a weak, pseudoirreversible inhibitor (74) of MAO while the corresponding non-fluorinated SAH hydrolase inhibitor (194) is a substrate for SAH hydrolase. Whereas replacement of one vinylic hydrogen atom for a chlorine atom (see compounds 191 and 195) has little effect on potency, a fluorine atom leads to increased potency of several orders of magnitude. For reasons which are not totally apparent, the isomer in which the fluorine atom is *trans* to the oxidized functionality, is optimal for potency (this is especially apparent in the series of MAO inhibitors; e.g., see compound 188 versus 192, Table 2). Difluoroolefins are the least potent of the fluoroolefin-containing inhibitors (MAO inhibitors) and approximate the activity seen with the chlorine-substituted analogs.

Little precise mechanistic studies have been undertaken with these inhibitors with the exception of the time-dependent inhibition of SAH hydrolase by 176. Stoichiometric loss of fluoride was observed by ^{19}F-NMR during the inactivation process. However, there is only circumstantial evidence to support the addition-elimination mechanism proposed in Figure 1; all attempts to isolate an enzyme fragment covalently bound to an inhibitor have so far been unsuccessful. If the rate-determining step in the enzyme inhibition process is an attack of an enzyme nucleophilic residue on a β-fluoro-α,β-unsaturated imine or ketone, kinetic analysis of addition-elimination reactions to similar systems indicate that the (*E*)-isomer is the more active isomer (73); this could explain in part the isomeric preference seen with MAO and SAH hydrolase inhibitors.

In conclusion, the approach utilizing a fluoroolefin functionality to design inhibitors of enzymes catalyzing an oxidation step has been successful in a number of cases. Further mechanistic studies are needed to confirm that the irreversible

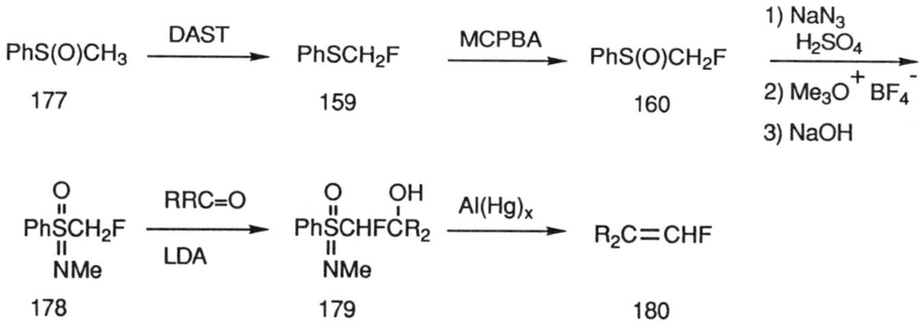

$$PhS(O)CH_3 \; (177) \xrightarrow{DAST} PhSCH_2F \; (159) \xrightarrow{MCPBA} PhS(O)CH_2F \; (160) \xrightarrow[\text{2) } Me_3O^+ BF_4^- \text{ 3) NaOH}]{\text{1) } NaN_3,\ H_2SO_4}$$

$$PhS(O)(=NMe)CH_2F \; (178) \xrightarrow[LDA]{RRC=O} PhS(O)(=NMe)CHFC(OH)R_2 \; (179) \xrightarrow{Al(Hg)_x} R_2C{=}CHF \; (180)$$

Scheme 51

$$HC(SPh)_2CH_2R \; (181) \xrightarrow[CH_3CN]{HgF_2} HC(SPh)(F)CH_2R \; (182) \xrightarrow[\text{2) HEAT}]{\text{1) MCPBA}} RCH{=}CHF \; (183)$$

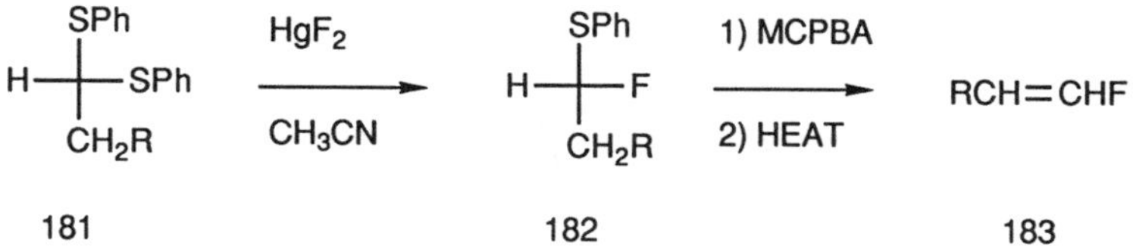

Scheme 52

TABLE 1. Fluoroolefin-containing mechanism-based enzyme inhibitors

ENZYME	EXAMPLES	REFERENCE
Monoamine oxidase	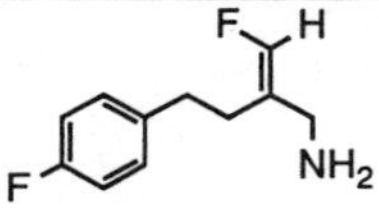 (184)	*71*
Monoamine oxidase (dual enzyme-activated)	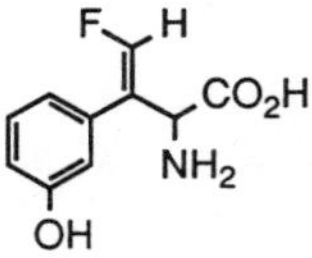 (185)	*62*
Polyamine oxidase	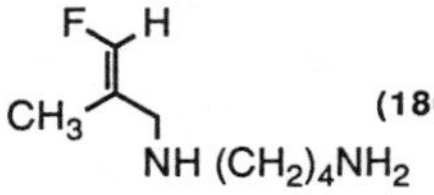 (186)	*47*
Diamine oxidase	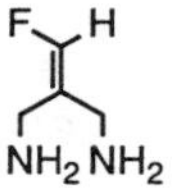 (187)	*47*
Semicarbazide-sensitive amine oxidase	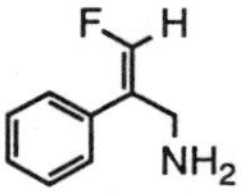 (188)	*72*
γ - Aminobutyric acid transaminase	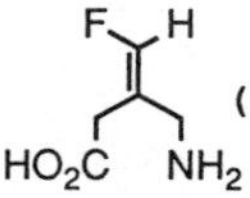 (189) 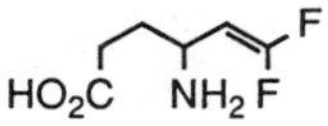 (190)	*47, 33*
S-Adensoylhomocysteine hydrolase	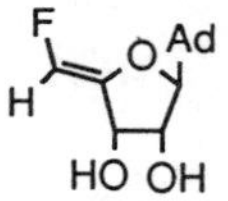 (176)	*68*

TABLE 2. Influence of fluorine atoms on relative inhibitory potency

ENZYME	BASIC STRUCTURE	X, Y = H	X = Cl Y = H	X = F Y = H	X = H Y = F	X, Y = F
Monoamine oxidase	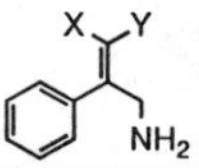	**190**	**191**	**188**	**192**	**193**
		IC_{50} = 400 μM	100 μM	0.018 μM	0.15 μM	10 μM
S-Adenosyl-homocysteine hydrolase	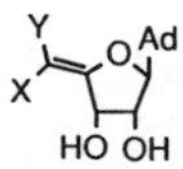	**194**	[X = H Y = Cl] **195**	**176**	**196**	
		K_I = (substrate)	1.6 μM	0.55 μM	1.04 μM	Not made

inhibition step is the addition of an active site enzyme nucleophile to the activated olefin followed by fluoride ion elimination.

Literature Cited

1. Silverman, R.B. *Mechanism-Based Enzyme Inactivation: Chemistry and Enzymology* CRC Press: Boca Raton, FL, **1988**; Vol. *I & II;*
2. Silverman, R.B. *Mechanism-Based Enzyme Inactivation: Chemistry and Enzymology;* CRC Press: Boca Raton, FL, **1988**; Vol. *II*; pp 107-117.
3. Schlosser, M. *Tetrahedron* **1978**, *34*, 3.
4. For an apparently comprehensive review of the preparation of fluorine-containing organic compounds see Takeuchi, Y. *Yuki Gosei Kagaku Kyokaishi* **1988**, *4*, 145.
5. Fuqua, S.A.; Duncan, W.G.; Silverstein, R.M. *J. Org. Chem.* **1965**, *30*, 1027.
6. Fuqua, S.A.; Duncan, W.G.; Silverstein, R.M. *J. Org. Chem.* **1965**, *30*, 2543.
7. Burton, D.J.; Herkes, F.E. *Tetrahedron Lett.* **1965**, 1883.
8. Herkes, F.E.; Burton, D.J. *J. Org. Chem.* **1967**, *32*, 1311.
9. Fuqua, S.A.; Duncan, W.G.; Silverstein, R.M. *Org. Synth., Coll. Vol. V.* **1973**, 390.
10. Herkes, F.E.; Burton, D.J. *Org. Synth. Coll. Vol. V*, **19** 949.
11. Schlosser, M.; Zimmerman, M. *Synthesis*, **1969**, 75.
12. Schlosser, M.; Zimmerman, M. *Synthesis*, **1971**, *104*, 2885.
13. Burton, D.J.; Greenlimb, P.E. *J. Org. Chem.* **1975**, *40*, 2796.
14. Hayashi, S.; Nakai, T.; Ishikawa, N.; Burton, D.J.; Naae, D.G.; Kesling, H.S. *Chem. Lett.* **1979**, *983*, and references cited therein.
15. Suda, M. *Tetrahedron Lett.* **1981**, *22*, 1421.
16. Fried, J.; Kittisopikul, S.; Halliman, E.A. *Tetrahedron Lett.* **1984**, *25*, 4329.
17. Motherwell, W.B.; Tozer, M.J.; Ross, B.C. *J. Chem. Soc., Chem. Commun.* **1989**, 1437.
18. Vinson, W.A.; Prickett, K.S.; Spahic, B.; Ortiz de Montellano, P.R. *J. Org. Chem.* **1983**, *48*, 4661.
19. *Chem. Pharm. Bull.* **1985**, *33*, 5137.
20. Burton, D.J.; Cox, D.G. *J. Am. Chem. Soc.* **1983**, *105*, 650.
21. Cox, D.G.; Gurusamy, N.; Burton, D.J. *J. Am. Chem. Soc.* **1985**, *107*, 2811.
22. Obayashi, M.; Ito, E.; Mutsui, K.; Kondo, M. *Tetrahedron Lett.* **1982**, *23*, 2323.
23. Obayashi, M.; Kondo, *Tetrahedron Lett.* **1982**, *23*, 2327.
24. Edwards, M.L.; Stemerick, D.M.; Jarvi, E.T.; Matthews, D.P.; McCarthy, J.R. *Tetrahedron Lett.* in press.
25. Blackburn, G.M.; Parratt, M.J. *J. Chem. Soc. Chem. Commun.*, **1983**, 886.
26. Fujita, M.; Hiyana, T. *Tetrahedron Lett.* **1986**, *27*, 3655.
27. McCarthy, J.R.; Matthews, D.P.; Edwards, M.L.; Stemerick, D.M.; Jarvi, E.T. *Tetrahedron Lett.* in press.
28. Inbasekaran, M.; Peet, N.P.; McCarthy, J.R.; LeTourneau, M.E. *J. Chem. Soc., Chem. Commun.* **1985**, 678.
29. Sauvetre, R.; Normant, J.F. *Tetrahedron Lett.* **1981**, *22*, 957.
30. Kolb, M.; Gerhart, F.; Francois, T.P. *Synthesis*, **1988**, 469.
31. Gillet, J.P.; Sauvetre, R.; Normant, J.F. *Synthesis* **1982**, 297.
32. Kende, A.S. *J. Org. Chem.* **1983**, *48*, 1384.

33. Kolb, M.; Barth, J.; Heydt, J.G.; Jung, M.J. *J. Med. Chem.* **1987**, *30*, 267.
34. Yang, Z.Y.; Burton, D.J. *Tetrahedron Lett.* **1990**, *31*, 1369.
35. Ishihara, T.; Yamana, M.; Ando, T. *Tetrahedron Lett.* **1983**, *24*, 5657.
36. Lee, S.H.; Schwartz, J. *J. Am. Chem. Soc.* **1986**, *108*, 2445.
37. Shen, T.Y.; Lucas, S.; Sarett, L.H. *Tetrahedron Lett.* **1961**, 43.
38. Bey, P.; Ducep, J.B.; Schirlin, D. *Tetrahedron Lett.* **1984**, *25*, 5657.
39. Bey, P.; Schirlin, D. *Tetrahedron Lett.* **1978**, 5225.
40. Bey, P. *In Enzyme-Activated Irreversible Inhibitors*; Seiler, N.; Jung, M.J.; Koch-Weser, J. (eds.); Elsevier/North-Holland Biomedical Press: Amsterdam **1978**; pp 27-41.
41. McDonald, I.A.; Lacoste, J.M.; Bey, P.; Palfreyman, M.G.; Zreika, M. *J. Med. Chem.* **1985**, *28*, 186.
42. Krapcho, A.P. *J. Org. Chem.* **1962**, *27*, 2375.
43. Kosuge, S.; Nakai, H.; Kurono, M. *Prostaglandins*, **1979**, 737.
44. Nishide, K.; Kobori, T.; Tunemoto D.; Kondo, K. *Heterocycles* **1987**, *26*, 633.
45. Tsushima, T.; Kawada, K. *Tetrahedron Lett.* **1985**, *26*, 2445.
46. Tsushima, T.; Kawada, K.; Ishihara, S.; Uchida, N.; Shiratori, O.; Higaki, J.; Hirata, M. *Tetrahedron* **1988**, *44*, 5375.
47. McDonald, I.A.; Bey, P. *Tetrahedron Lett.* **1985**, *26*, 3807.
48. Rico, I.; Cantacuzene, D.; Wakselman, C. *Tetrahedron Lett.* **1981**, *22*, 3405.
49. Leroy, J.; Molines, H.; Wakselman, C. *J. Org. Chem.* **1987**, *52*, 290.
50. Leroy, J. *J. Org. Chem.* **1981**, *46*, 206.
51. Castelhano, A.L.; Krantz, A. *J. Am. Chem. Soc.* **1987**, *109*, 3491.
52. Nguyen, T.; Wakselman, C. *J. Org. Chem.* **1989**, *54*, 5640.
53. Fuchikami, T.; Shibata, Y.; Suzuki, Y. *Tetrahedron Lett.* **1986**, *27*, 3173.
54. Kendrick, D.A.; Kolb, M. *J. Fluorine Chem.* **1989**, *45*, 265.
55. Ichikawa, J.; Sonoda, T.; Kobayashi, H. *Tetrahedron Lett.* **1989**, *30*, 1641.
56. Metcalf, B.W.; Jarvi, E.T.; Burkhart, J.P. *Tetrahedron Lett.*, **1985**, *26*, 2861.
57. Greuter, H.; Lang, R.W.; Romann, A.J. *Tetrahedron Lett.* **1988**, *27*, 3291.
58. Nakai, T.; Tanaka, K.; Ogasawara, K.; Ishikawa, N. *Chem. Lett.* **1981**, 1289.
59. Kitagawa, O.; Taguchi, T.; Kobayashi, Y. *Tetrahedron Lett.* **1988**, *29*, 1803.
60. Taguchi, T.; Kitagawa, O.; Suda, Y.; Ohkawa, S.; Hashimoto, A.; Iitaka, Y.; Kobayashi, Y. *Tetrahedron Lett.* **1988**, *29*, 5291.
61. Yamana, M.; Ishihata, T.; Ando, T. *Tetrahedron Lett.* 1983, *24*, 507.
62. McDonald, I.A.; Lacoste, J.M.; Bey, P.; Wagner, J.; Zreika, M.; Palfreyman, M.G. *Bioorg. Chem.* **1986**, *14*, 103.
63. McDonald, I.A.; Palfreyman, M.G.; Jung, M.; Bey, P. *Tetrahedron Lett.* **1985**, *26*, 4091.
64. Reutrakul, V.; Rukachaisirikul, V. *Tetrahedron Lett.* **1983**, *24*, 725.
65. McCarthy, J.R.; Peet, N.P.; LeTourneau, M.E.; Inbasekaran, M. *J. Am. Chem. Soc.* **1985**, *107*, 735.
66. McCarthy, J.R.; Matthews. D.P.; Edwards, M.L.; Stemerick, D.M.; Jarvi, E.T. *Tetrahedron Lett.* in press.

67. Umemoto, T.; Tomizawa, G. *Bull. Chem. Soc. Jpn.* **1986**, *59*, 3625.
68. McCarthy, J.R.; Jarvi, E.T.; Matthews, D.P.; Edwards, M.L. Prakash, N.J.; Bowlin, T.L.; Mehdi, S.; Sunkara, P.S.; Bey, P. *J. Am. Chem. Soc.* **1989**, *111*, 1127.
69. Boys, M.L.; Collinton, E.W.; Finch, H.; Swanson, S.; Whitehead, J.F. *Tetrahedron Lett.* **1988**, *29*, 3365.
70. Purrington, S.T.; Pittman, J.H. *Tetrahedron Lett.* **1987** *28*, 3901.
71. Zreika, M.; Fozard, J.R.; Dudley, M.W.; Bey, P.; McDonald, I.A.; Palfreyman, M.G. *J. Neural. Transm.* **1989**,*1*, 243.
72. Lyles, G.A.; Marshall, C.M.S.; McDonald, I.A.; Bey, P.; Palfreyman, M.G. *Biochem. Pharmacol.* **1987**, *36*, 2847.
73. Modena, G. *Acc. Chem. Res.* **1971**, *4*, 73.
74. For a discussion on pseudoirreversible inhibition see Rando, R.R.; Eigner, A. *Molec. Pharmacol.* **1977**, *13*, 1005.
75. Bergstrom, D.E.; Ng, M.W; Wong, J.J. *J. Org. Chem.* **1983**, *48*, 1902.
76. Kitazume, T.; Ohnogi, T.; Miyauchi, H.; Yamazaki, T.; Watanabe, S. *J. Org. Chem.* **1989**, *54*, 5630.

RECEIVED October 16, 1990

BIOLOGICAL APPLICATIONS

Chapter 9

Fluorine-Substituted Neuroactive Amines

Kenneth L. Kirk

Laboratory of Bioorganic Chemistry, National Institute of Diabetes and Digestive and Kidney Diseases, National Institutes of Health, Bethesda, MD 20892

Fluorine substitution on the aromatic ring of catecholamines has a striking effect on their selectivities for the α- and β-subtypes of adrenergic receptors, with potency at a given receptor dependent on the site of fluorine substitution. The results of testing of new analogs synthesized to probe mechanisms of adrenergic selectivity indicate that a direct effect of the C-F bond on agonist-receptor interaction may be more important than an indirect effect of the C-F bond on the conformation of the ethanolamine side-chain. Research on the effect of fluorine substitution on biological activities of neuroactive amines has been extended to include examination of adrenergic antagonists.

The naturally occurring catecholamines dopamine (**1**), norepinephrine (**2**), and epinephrine(**3**) (Figure 1) play key roles in neurotransmission, metabolism, and in the control of various physiological processes. For example, norepinephrine is the primary neurotransmitter in the sympathetic nervous system and also functions as a neurotransmitter in the central nervous system. Epinephrine, elaborated by the adrenal gland, has potent effects on the heart, vascular and other smooth muscles. Dopamine is an important neurotransmitter in the central nervous system, and has important peripheral effects in such organs as the kidney and heart. The importance of these effects has made the search for drugs that can mimic, inhibit, or otherwise modulate the effects of these catecholamines an important area of medicinal chemistry.

In designing analogs of biologically important molecules, the replacement of a carbon-hydrogen bond with a carbon-fluorine bond has become an important strategy. The relatively small steric alterations

Published 1991 American Chemical Society

that result from this substitution can facilitate interactions of the analog with a variety of biological systems, such as enzyme active sites, receptor systems, and transport systems. On the other hand, the altered electronic properties of the analog, and/or altered available reaction pathways, can change the biological properties of the analog dramatically. Accordingly, the synthesis and biological evaluation of fluorinated compounds has developed into a major area of medicinal chemistry, and these efforts have led to a host of useful medicinal agents and pharmacological tools. In addition, incorporation of the positron emitting isotope ^{18}F ($t_{1/2}$ = 110 min) has proven to be an effective approach to the development of scanning agents for use in positron emission tomography (PET).

Using a variety of synthetic methodologies, including procedures developed in our laboratory, we have prepared a series of biogenic amines, including adrenergic agonists, with fluorine substituted at various aromatic ring positions. Fluorine substitution markedly affects the biological properties of these amines, a fact that has provided several useful pharmacological tools. Altered receptor affinities of fluorinated adrenergic agonists has revealed considerable information on the mode of interaction of adrenergic agonists with receptors. Elucidation of the pharmacological behavior of these analogs also has provided important information that has facilitated the development of ^{18}F-labelled analogs as PET scanning agents. In this report, I will provide an updated review of our research in this area, and describe future plans for this research.

Background

Our extensive efforts to prepare previously unknown ring-fluorinated imidazoles culminated in 1970 in the development of a photochemical variant of the Schiemann reaction that led to facile syntheses of fluoroimidazoles (*1-3*). Examples of analogs we prepared included 4- and 2-fluoro-L-histidine (**4a,b**), 4- and 2-fluorohistamine (**5a,b**), and 4-fluoroimidazole-5-carboxamide riboside (**6**) (Figure 2). These fluorinated analogs possessed useful biological properties, with the position of the fluorine substitution having a dramatic effect on biological activity, a phenomenon that becomes central to later work. For example, 2-fluoro-L-histidine, but not 4-fluoro-L-histidine, is incorporated into bacterial and mammalian protein in vivo, and has potent antiviral and antibacterial activity. These and other biological properties of fluoroimidazoles have been reviewed (*4*).

While the precise mechanisms of action of many of the biological actions of fluoroimidazoles have not been elucidated, the dramatic lowering of the imidazole nitrogen pK_a caused by the presence of fluorine on an adjacent carbon presumably is reflected in certain of these properties. Consideration of other biologically critical ionizable groups that might be altered by the electronegative effects of fluorine substitution led us to realize that, despite the presence of the catechol

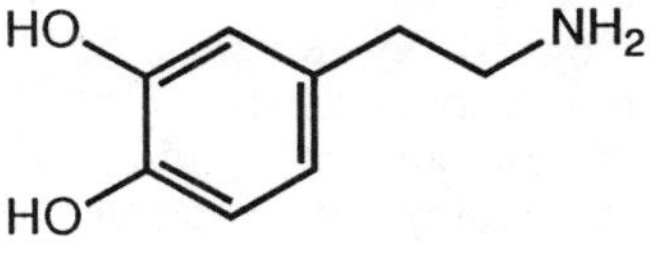

1

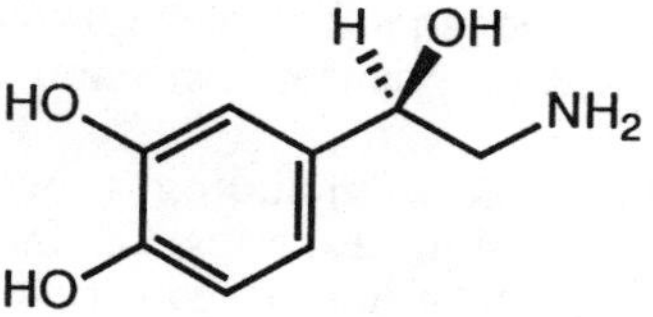

2

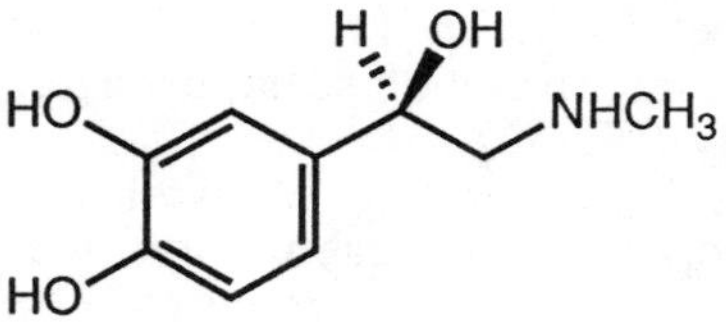

3

Figure 1: Naturally occurring catecholamines: Dopamine (**1**), norepinephrine (**2**), and epinephrine (**3**).

4a: R_1 = F; R_2 = H
4b: R_1 = H; R_2 = F

5a: R_1 = F; R_2 = H
5b: R_1 = H; R_2 = F

6

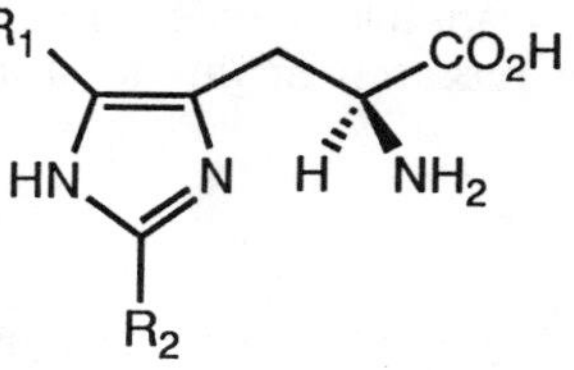

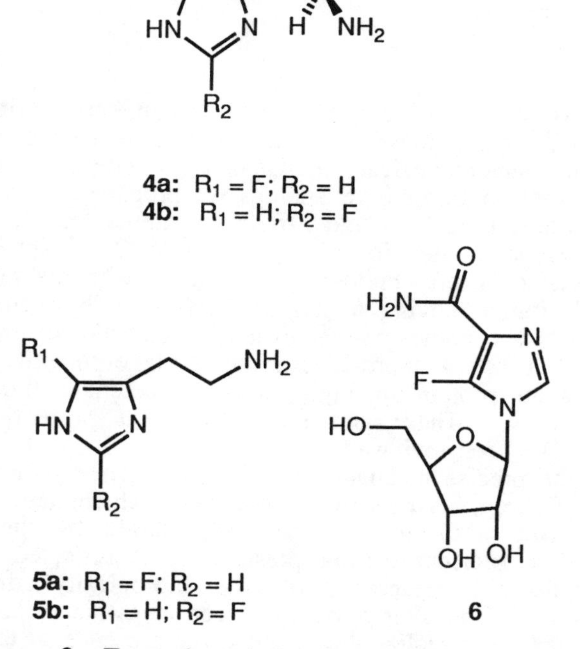

Figure 2: Examples of ring-fluorinated imidazoles.

ring in an important series of biogenic amines, and the fact that a fluorine-induced increase in phenolic pK_a would be expected, and is documented, there were no published reports of ring-fluorinated biogenic catecholamines.

Fluorinated Dopamines

With the above factors in mind, we synthesized 2-, 5-, and 6-fluorodopamine (**7a-c**) (Figure 3) (*5*) in our initial study of the effects of ring-fluorination on the biological properties of biogenic amines. While significant differences were observed in certain systems, fluorine substitution did not have profound effects on the interaction of dopamine with dopamine receptors, or with adrenergic receptors (*6,7,8*). In fact, the qualitatively similar pharmacological behavior of 6-fluorodopamine and dopamine is a useful property, in that this adds validity to the use of [^{18}F]-6-fluoro-DOPA as a biological precursor of [^{18}F]-6-fluorodopamine in PET scanning of central dopaminergic function (*9*). The utility of [^{18}F]-6-fluoro-DOPA as a PET-scanning agent is further increased by the fact that fluorine in the 6-position retards the rate of metabolism of both 6-fluoro-DOPA and 6-fluorodopamine by catechol-*O*-methyltransferase, an important catechol-metabolizing enzyme (*10,11*). Recently, the in vivo conversion [^{18}F]-6-fluorodopamine to [^{18}F]-6-fluoronorepinephrine has been exploited in a procedure developed for PET imaging of adrenergic innervation in the heart (*12*).

The Effects of Ring-Fluorination on the Activities of Adrenergic Agonists

Norepinephrine (NE) controls a host of physiological processes, including such fundamentally important functions as blood pressure, heart rate, smooth muscle control, and metabolism of liver, fat cells, adrenal glands, pancreas, etc. These actions are mediated through the interaction of norepinephrine with adrenergic receptors. The response of a tissue to the presence of norepinephrine, or other adrenergic agonists, depends on the presence and type of adrenergic receptor present in the tissue. The discovery of different types of adrenergic receptors [termed α- and β-adrenergic receptors (these are further divided into subtypes, α_1, α_2, β_1, β_2, etc.)] in tissues has been the basis for the development of many clinically useful drugs for the treatment of a variety of illness, including such serious problems as hypertension and asthma.

Adrenergic Selectivities of Ring-Fluorinated Norepinephrines. 2-, 5-, and 6-fluoronorepinephrine (FNE) (**8a-c**) (Figure 3) were synthesized from the corresponding fluorinated veratraldehydes (*13,14*). Examination of the potencies of 2-, 5-, and 6-FNE in stimulating the contraction of isolated guinea pig aorta (α_1-response) revealed that, while NE and 5- and 6-FNE had comparable activity, 2-FNE showed no significant activity. In contrast, with respect to increasing the rate of beating of

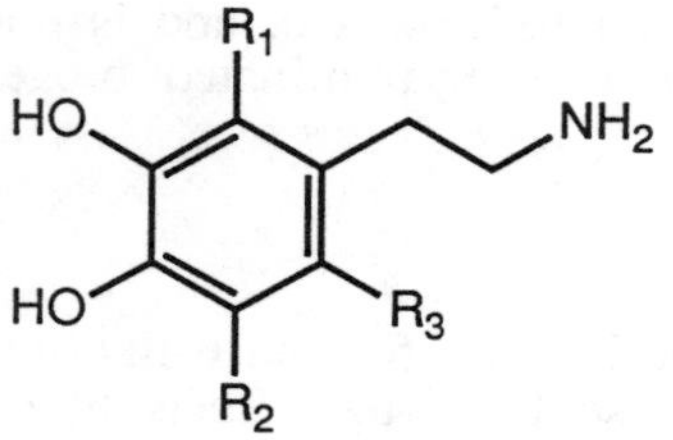

7a: $R_1 = F, R_2 = R_3 = H$
b: $R_2 = F, R_1 = R_3 = H$
c: $R_3 = F, R_1 = R_2 = H$

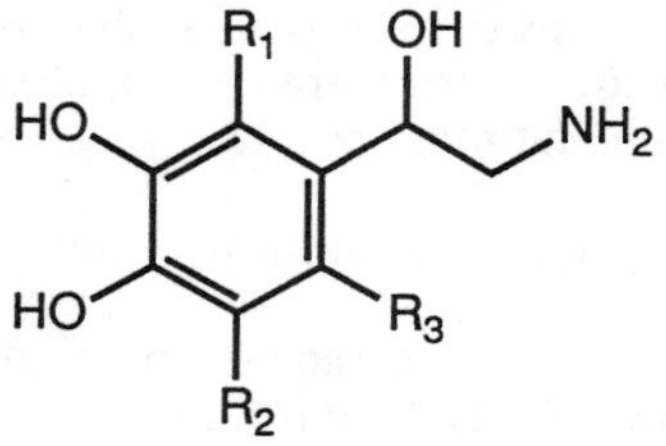

8a: $R_1 = F, R_2 = R_3 = H$
b: $R_2 = F, R_1 = R_3 = H$
c: $R_3 = F, R_1 = R_2 = H$

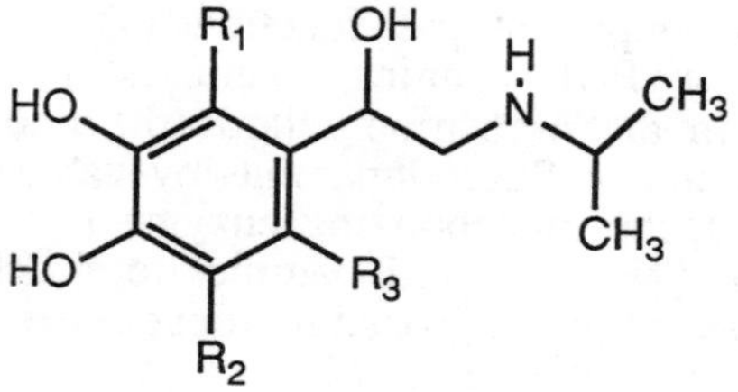

9a: $R_1 = F, R_2 = R_3 = H$
b: $R_2 = F, R_1 = R_3 = H$
c: $R_3 = F, R_1 = R_2 = H$

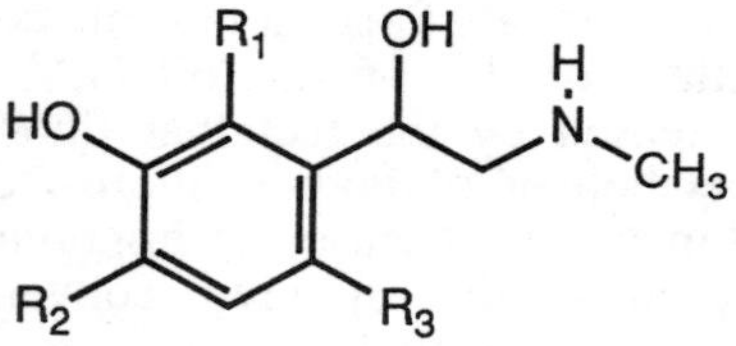

10a: $R_1 = F, R_2 = R_3 = H$
b: $R_2 = F, R_1 = R_3 = H$
c: $R_3 = F, R_1 = R_2 = H$

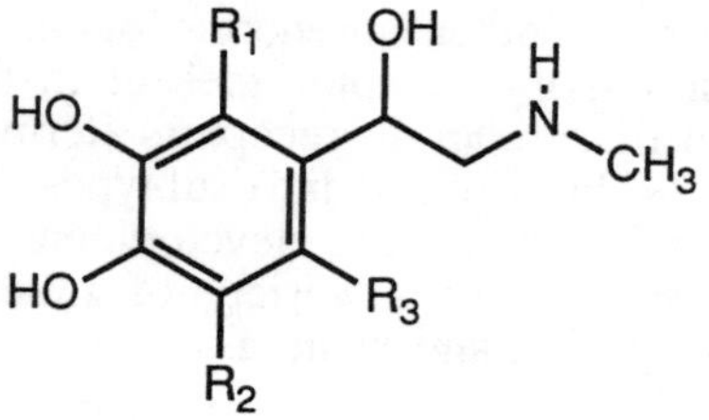

11a: $R_1 = F, R_2 = R_3 = H$
b: $R_2 = F, R_1 = R_3 = H$
c: $R_3 = F, R_1 = R_2 = H$

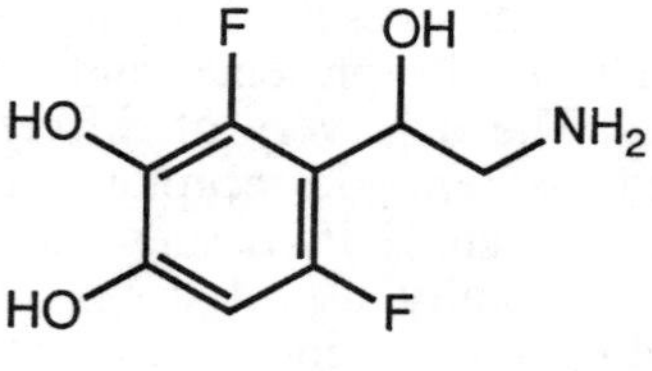

12

Figure 3: Ring-fluorinated neuroactive amines: F-DA (**7**); F-NE (**8**); F-ISO (**9**); F-PE (**10**); F-EPI (**11**); 2,6-DiF-NE (**12**).

isolated guinea pig atria (β_1-response), 5-FNE was two-fold more potent than NE and 2-FNE, while 6-FNE was 100-fold less potent. Subsequent receptor binding assays related these differences in agonist responses to receptor affinities, and confirmed that 2-FNE is a selective β-adrenergic agonist, while 6-FNE is a selective α-adrenergic agonist. 5-FNE behaved comparably to NE, although, as noted above, in certain systems, a somewhat higher activity is seen with 5-FNE.

These initial results subsequently were confirmed in many additional systems. Of particular significance are the close structural similarities of the fluorinated analogs to NE, a fact that has made these analogs very useful pharmacological tools. For example, 2- and 6-FNE were used to map regional differences in concentrations of α- and β-adrenergic receptors in various parts of rat brain (*15*). In another example, synergism between α- and β-adrenergic receptors in pineal gland was revealed using these analogs as selective agonists (*16*). These, and numerous other studies have demonstrated the value of these analogs. In addition, 2- and 6-FNE both have been shown to function as "false transmitters," that is, they are taken up into adrenergic neurons, displacing endogenous norepinephrine in the process (*17*). The uptake and storage of [^{18}F]-6-FNE, synthesized in vivo from [^{18}F]-6-F-dopamine, is the basis for visualization of peripheral adrenergic innervation by PET imaging (*12*).

Effect of Fluorine Substitution on Isoproterenol and Phenylephrine. Norepinephrine is a natural "mixed" adrenergic agonist--that is, it is active at both α- and β-adrenergic receptors. To explore the generality of fluorine-induced adrenergic selectivities, we extended our studies to include fluorinated analogs of the potent and selective β-adrenergic agonist, isoproterenol (ISO) (**9a-c**) (Figure 3) and the selective α-adrenergic agonist, phenylephrine (PE) (**10a-c**) (*18,19*) (Figure 3). In the ISO series, a "negative" feature of the influence of fluorine was indicated by the fact that 6-F-ISO had markedly reduced activity as a β-adrenergic agonist, but no apparent α-adrenergic activity was induced. 2-F-ISO and 5-F-ISO had β-adrenergic activity comparable to ISO. However, in the PE series, we observed the first indication that fluorine could increase potency as well as induce selectivity. For example, the rank order of α_1-adrenergic potency (stimulation of contraction of isolated guinea pig arota) was 6-F-PE > PE = 4-F-PE >> 2-F-PE.

Adrenergic Activity of Fluorinated Epinephrines. While epinephrine (EPI), as NE, is a naturally occurring "mixed" agonist, the relative potencies at α- and β-adrenergic receptors differ between the two. At α-receptors, the rank order of potency for a series of agonists is EPI > NE > ISO while at β-receptors the order is ISO > NE > EPI. Fluorinated analogs of EPI (**11a-c**) (Figure 3) were synthesized and shown to have the same impressive adrenergic selectivities as seen with F-NEs (*20*). However, as in the PE series, fluorine was shown also to increase potencies. Thus, at β_1- and β_2-adrenergic receptors the rank order of potency was 2-F-EPI > EPI >> 6-F-EPI and at α-adrenergic receptors, the rank order of potency was 6-F-EPI > EPI >> 2-F-EPI. The combined selectivity and increased potencies of 2- and 6-F-EPI give these analogs increased potential as agents for receptor characterization, as studies in progress have confirmed (V. Doze, Stanford University, personal communication, 1990).

Adrenergic Activity of 2,6-Difluoronorepinephrine. To explore further whether adrenergic selectivities are based primarily on selective inhibition of binding, we prepared 2,6-difluoronorepinephrine (**12**) (*21*) (Figure 3). With fluorine at both receptor-determinant positions, this analog was designed to determine if the effect of the two fluorines would cancel (to produce an active non-selective analog) or would be additive (to produce an analog with reduced affinity at both receptor types). The rank order of affinities at α- and β-adrenergic receptors suggest that the latter analysis pertains (α: NE = 6-FNE > 2,6-DiFNE > 2-FNE; β: NE = 2-FNE > 2,6-DiFNE > 6-FNE) (Kirk, K. L.; Chen, G.; Daly, J. W.; Gusovsky, F.; Creveling, C. R., NIH, unpublished data).

Studies on the Mechanism of Fluorine-Induced Adrenergic Selectivities

A summary of the effects of fluorine substitution on the adrenergic selectivities of adrenergic agonists is given if Figure 4. As can be seen, the absence of fluorine in position 2 is required for potent activity at the α-adrenergic receptor, while the absence of fluorine in position 6 is required for potent activity at the β-adrenergic receptor. A striking feature of these results is the "anti-symmetric" nature of the phenomenon. Thus, in addition to providing useful pharmacological tools, the observed selectivities appear to reflect a significant, if subtle, difference in binding modes of α- and β-adrenergic agonists. One concept that prompted initiation of these studies was the expectation that altered biological properties could be interpreted based on predictable changes in physico-chemical properties of the analog produced by fluorination of the catechol ring. As will be shown in this section, we are now using the results of biological testing to attempt to define just what these effects of fluorine substitution are.

Conformational Effects of Fluorine Substitution. Mechanisms considered to explain adrenergic selectivities of fluorinated norepinephrine (and related adrenergic agonists) have included: 1) an indirect effect of the C-F bond on the conformation of the ethanolamine side-chain or 2) a direct effect of the C-F bond on agonist-receptor interaction. In the first formulation, proposals were made that fluorine situated in a position ortho (position 2 or 6) to the ethanolamine side chain creates a bias for side chain conformations favorable for binding to β- and α-adrenergic receptors, respectively.

Intramolecular Hydrogen Bonding. Putative conformational bias was attributed initially to intramolecular hydrogen bonding between the benzylic OH group and the 2- or 6-fluoro substituent to produce conformations recognized by the β- and α-adrenergic receptor, respectively, as shown in Figure 5 (*13*). The absence of precedent for such an interaction, the likely instability of such a hydrogen bond, and a knowledge of preferred agonist conformation (see below) caused us to abandon this proposal.

Electrostatic Repulsion Between the Benzylic OH and Fluorine. A series of β-aminotetralols [for example, the semi-rigid analog (**13**) (Figure 6) of ISO (*22*)] have been shown to possess potent β-adrenergic activity. A comparison of **13** with 2-F-NE (also a β-adrenergic agonist) shows that a similar juxtaposition of the functional groups requires that the benzylic OH group and the ortho-fluorine substituent be oriented away from each other. This prompted us to consider the possibility that an electrostatic repulsion between the dipoles of the C-F bond and the C-OH bond could influence conformational preferences of the fluorinated analogs such that high populations of conformers could produce selective adrenergic agonists, as shown in Figure 7. (*19*). Support for this proposal was found in the published report that, in a polycyclic system, conformational effects of a similar interaction were observed directly using NMR (*23*).

The same electronic repulsive interaction was proposed independently by DeBernardis, who used this concept to design a series of α_2-selective adrenergic agonists (Figure 8) (*24*). According to the "electrostatic repulsion based conformational prototype" (ERBCOP) concept, α_2-selectivities exhibited by semi-rigid analogs such as **14** and by 6-FNE are related in that each compound is constrained (either by the bicyclic structure or by electronic repulsion) to be in similar conformations. However, our subsequent observation that 6-F-EPI relative to EPI has reduced α_2- vs. α_1-potency suggests that this relationship may not be valid, since such a reduced α_2- vs. α_1-potency is not consonant with results obtained by DeBernardis with semirigid analogs (*20*).

Fluorine-Induced Adrenergic Selectivity in a Conformationally Constrained Analog. We recently have obtained negative evidence that conformational factors may not be important in fluorine-induced adrenergic selectivities. As noted above, a series of β-amino tetralols are potent β-adrenergic agonists (*22,25*). We have synthesized the β-aminotetralol **15** and the fluorinated analog **16** (Figure 9) and compared the β-adrenergic activity of the two. In this analog, the fluorine substituent formally is in the "6"-position. Thus, **16** is constrained to be in a conformation favorable for interaction with the β-adrenergic receptor, but has fluorine in the "6"-position, a feature shown to affect adversely β-adrenergic activity. The nearly 400-fold decrease in affinity of **16** relative to **15** at β-adrenergic receptors indicates that the effect of fluorine on receptor affinity is not related to conformational factors in

Compound	α	β
Norepinephrine	+	+
2-F	-	+
5-F	+	+
6-F	+	-
Isoproterenol	-	+
2-F	-	+
5-F	-	+
6-F	-	-
Phenylephrine	+	-
2-F	-	+
4-F	+	-
6-F	++	--
Epinephrine	+	+
2-F	-	++
6-F	++	-

Figure 4: A summary of the selectivities of fluorinated adrenergic agonists for α- and β-adrenergic agonists.

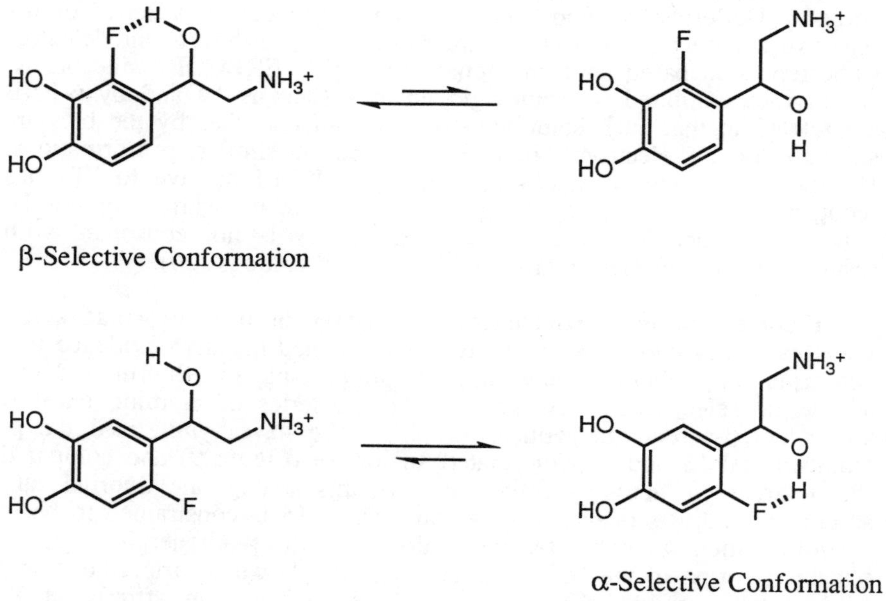

β-Selective Conformation

α-Selective Conformation

Figure 5: Hydrogen bonding between the benzylic hydroxyl group and an ortho-situated fluorine would favor opposite conformations for 2- and 6-F-NE.

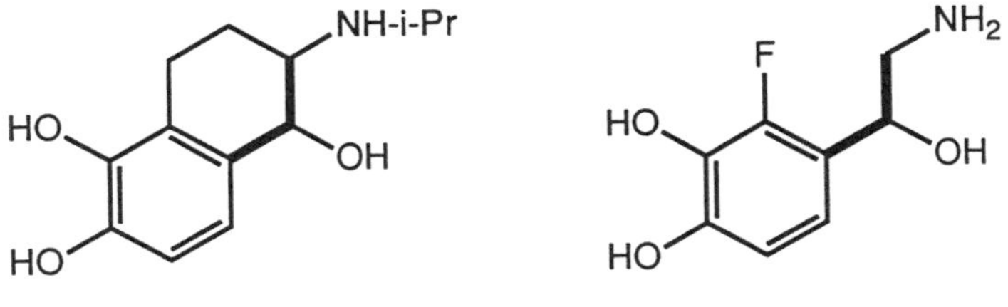

13

Figure 6: A semi-rigid analog, **13**, of isoproterenol is a potent β-adrenergic agonist. A comparable conformation of 2-F-NE has the hydroxyl group oriented away from fluorine.

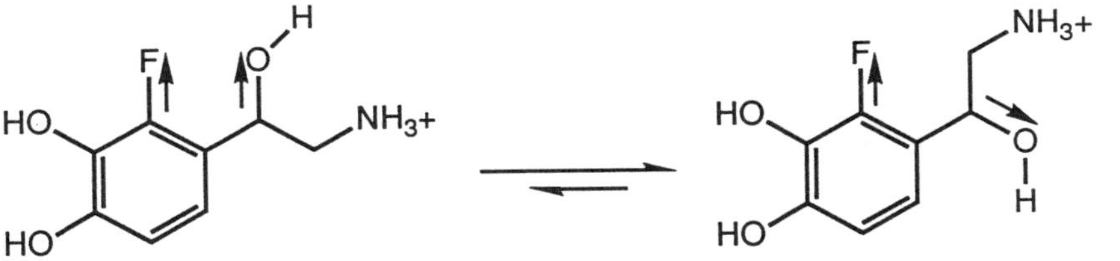

β-Selective Conformation

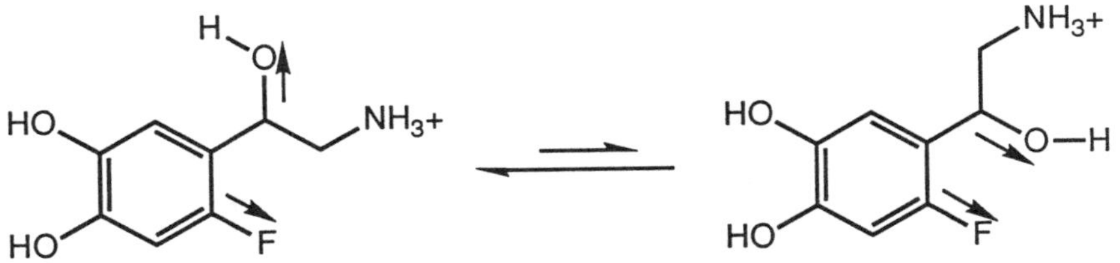

α-Selective Conformation

Figure 7: Electrostatic repulsion between the benzylic hydroxyl group and an ortho-situated fluorine would favor conformations opposite to those predicted for hydrogen bonding, but would be consistent with the adrenergic activities of semi-rigid analogs.

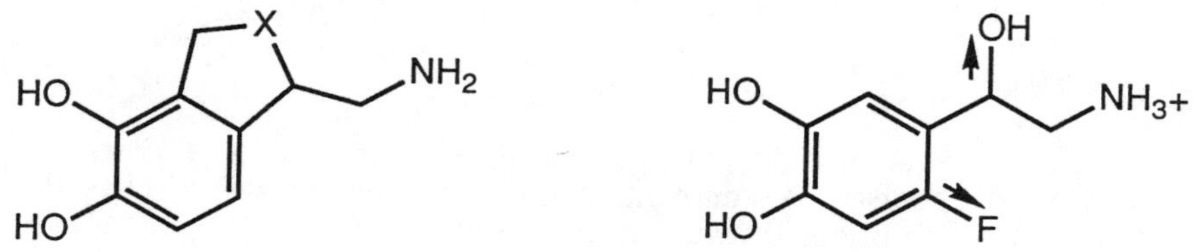

"6-ERBCOPS"

14: X = CH_2; O

Figure 8: α_2-Selective semi-rigid analogs (**14**) have the same conformation that would be predicted by electrostatic repulsion in 6-F-NE.

15 **16**

Figure 9: A fluorinated analog (**16**) of the semi-rigid β-aminotetralol (**15**).

this sterically constrained analog (Kirk, K. L.; Calderon, S.; Daly, J. W.; Gusovsky, F.; Creveling, C. R., NIH, unpublished data).

Fluorine-Induced Electronic Effects and Adrenergic Selectivities. It is evident that the presence of fluorine will alter the electronic distribution of the catechol ring. As an alternative explanation for fluorine-induced adrenergic selectivities, we have considered the possibility that electronic perturbations caused by the presence of the highly electronegative fluorine substituent may alter specific interactions of the aromatic ring with charged sites on the receptor protein. The fact that selectivities appeared to result from a selective inhibition of binding of the fluorinated agonist to α- or β-adrenergic receptors led us to propose that the greatly decreased electron density associated with the carbon to which fluorine is bound might block interaction of that ring position with a positive charge on the receptor protein. To explain the observed selectivities, we proposed that the β-adrenergic receptor would interact regiospecifically with the agonist 6-position while the α-adrenergic receptor would interact at the agonist 2-position, as shown schematically in Figure 10 (*18*).

Kocjan et al. (*26*) proposed an alternative mechanism by which the presence of a C-F bond could alter receptor binding selectively. According to their proposal, the negative end of the C-F dipole would be repulsed by a negative charge on the receptor protein, this charge being present at different positions for α- and β-adrenergic receptors (Figure 11).

Using photochemically induced dynamic nuclear polarization (CIDNP), Muzskat (*27*) measured the electronic distributions in the highest occupied molecular orbitals (HOMOs) of NE and its ring fluorinated analogs. Of particular significance was his demonstration that the electronic distribution of the HOMO of 6-FNE is different from the other three compounds (Figure 12). According to his analysis, fluorine substituted at C-6 will stabilize Ψ_{1a} to a greater extent than Ψ_{1s}, leading to an inversion of the relative stabilities of the two orbitals. If the catechol ring is involved in charge-transfer interactions with a protein aromatic amino acid, changes in frontier orbital characteristics could be very important in defining receptor selectivities. This issue is being pursued.

Is the Benzylic Hydroxyl Group Required for Adrenergic Selectivity? A fundamental difference between mechanisms mediated by conformational factors, and those mediated directly by changes in electronic characteristics of the aromatic ring is that the former requires specific interactions with the alkylamine side chain. To explore further the effect of altering the side-chain structure on fluorine-induced adrenergic selectivities, we prepared fluorinated analogs (**17b,c**) of 1-(3,4-dihydroxyphenyl)-3-t-butylamino-2-propanol (**17a**) (Figure 13), a potent β-adrenergic agonist not possessing a benzylic hydroxyl group (*28*). The fact that the 6-fluoro-analog had dramatically lower β-adrenergic

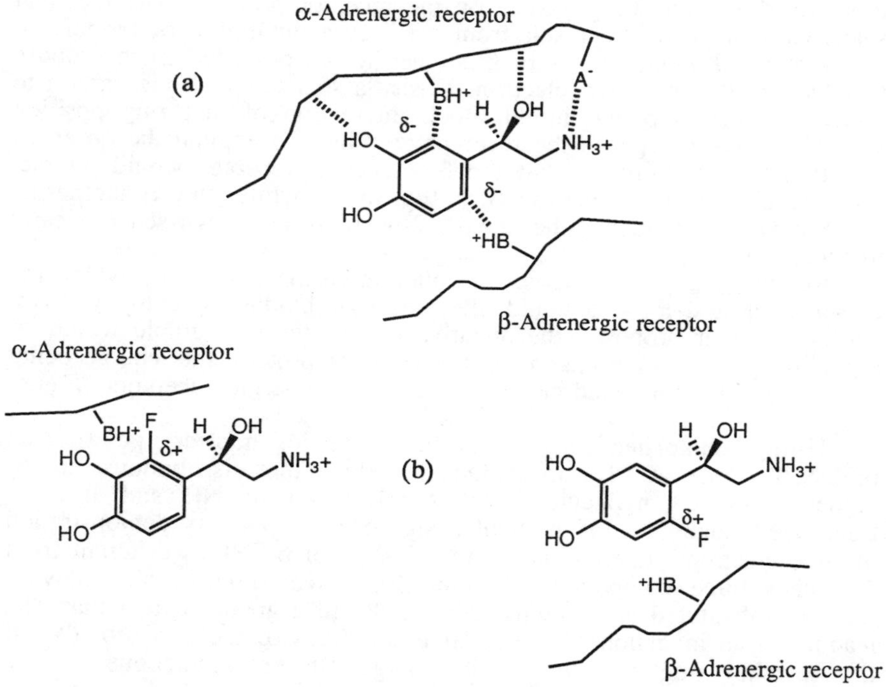

(a)

α-Adrenergic receptor

BH+ H OH A⁻ δ- HO NH3+ δ- HO +HB

β-Adrenergic receptor

α-Adrenergic receptor

BH+ F H OH δ+ HO NH3+ HO

(b)

H OH HO NH3+ δ+ HO F +HB

β-Adrenergic receptor

Figure 10: (a) Interaction of a putative positive charge on the aromatic ring was proposed as part of the binding interactions. This interaction differs for *α*- and *β*-adrenergic receptors. (b) The proposal was made that the decreased electron density on the carbon bearing the fluorine substituent inhibits binding to the receptor.

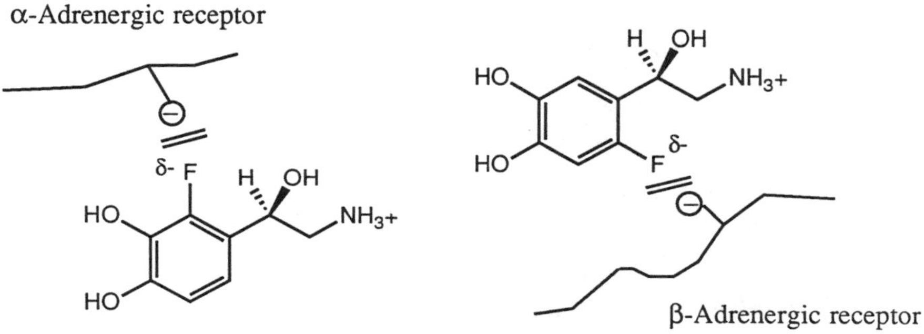

α-Adrenergic receptor

β-Adrenergic receptor

Figure 11: An alternative proposal involves the repulsive interaction of the negative end of the C-F dipole with a negative charge on the adrenergic receptor, with this charge having different locations on the α- and β-adrenergic receptor.

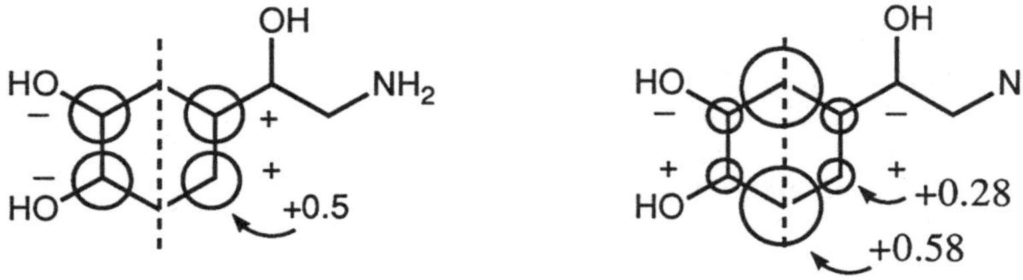

ψ_{1A} — NE, 2-FNE, 5-FNE ψ_{1S} — 6-FNE

Figure 12: The electronic distribution of the highest occupied molecular orbitals of NE, 2-, 5-, and 6-F-NE as determined by photochemically induced dynamic nuclear polarization experiments (adapted from Muszkat, 1988, ref. 25).

activity, while the corresponding 2-fluoro analog had comparable, or greater, potency relative to the parent suggests that the mechanism(s) for selectivity is not dependent on interactions with the side chain, and may well be related to fluorine-induced alterations in the interaction of the aromatic ring with receptor amino acid residues. These results also suggest that, on binding to the β-adrenergic receptor, **17** adopts a conformation such that the 2- and 6-positions of the aromatic ring coincide with the 2- and 6-positions of phenethanolamine β-adrenergic agonists (*29*).

Does Fluorine Substitution also Affect Binding of Adrenergic Antagonists? The phenoxypropanolamine moiety present in **17** is a common structural feature of many β-adrenergic antagonists. The observation of greatly decreased affinity of **17c** for β-adrenergic receptors gives credence to the study of the effects of fluorine on related antagonists. To this end, we prepared fluorinated analogs (**18b,c**) of the potent β_1-adrenergic antagonists practolol (**18a**) (Figure 14). Initial results suggest that fluorine has no effect on affinities in this series. Other fluorinated antagonists will be examined, particularly those having more clearly defined "2"- and "6"-positions (Kirk, K. L.; Padgett, W. Unpublished observations).

Specific Sites for Ligand-Receptor Interactions

The recognition that the effect of fluorine on the binding of amines to adrenergic receptors apparently results from electronic perturbations of the fluorinated aromatic ring has prompted us to examine possible aromatic ring--receptor protein interactions that might be affected by these perturbations. In this regard, the recent advances in the cloning and sequence analyses of neurotransmitter receptor genes has opened the way for studies of structure/function relationships to a degree that previously had been impractical. Adrenergic receptors belong to a family of receptor glycoproteins that mediate their actions by interaction with guanine nucleotide binding regulatory proteins. These receptor proteins show a high degree of homology and are linked evolutionarily. Besides adrenergic receptors, other important examples are muscarinic receptors and the visual protein, opsin. These proteins are characterized by having an extracellular amino terminus, seven membrane-spanning loops, and the carboxy terminus region located in the cytoplasm (shown schematically in figure 15).

The ability to produce adrenergic receptors with single amino acid substitutions at sites suspected to be involved with binding and/or agonist response already has provided much information on the structural requirements for ligand binding (*30*). As stressed above, substitution of fluorine on the aromatic ring of NE or EPI introduces only minor steric changes into the molecule. When the marked influence on adrenergic selectivities this substitution produces is considered, these fluorinated analogs should have unique advantages as probes for struct-

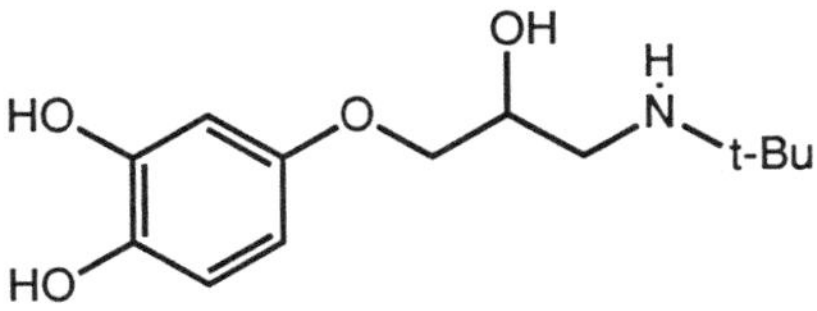

OH
H
HO O N t-Bu
HO

17a

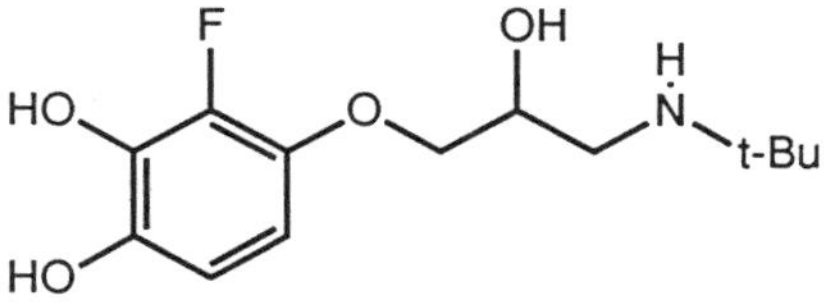

F OH
H
HO O N t-Bu
HO

17b

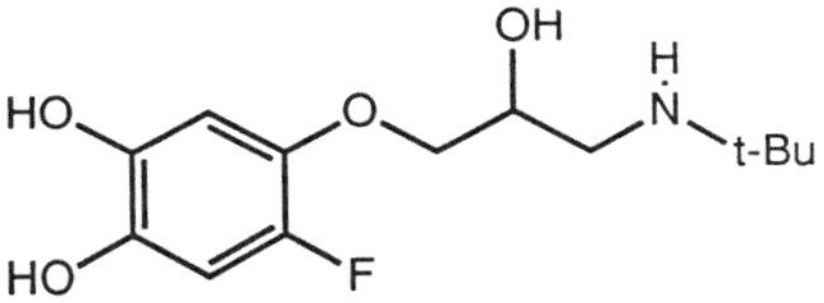

OH
H
HO O N t-Bu
HO F

17c

Figure 13. Fluorinated analogs (**17b,c**) of the β-adrenergic agonist **17a** that do not contain the ethanolamine side chain.

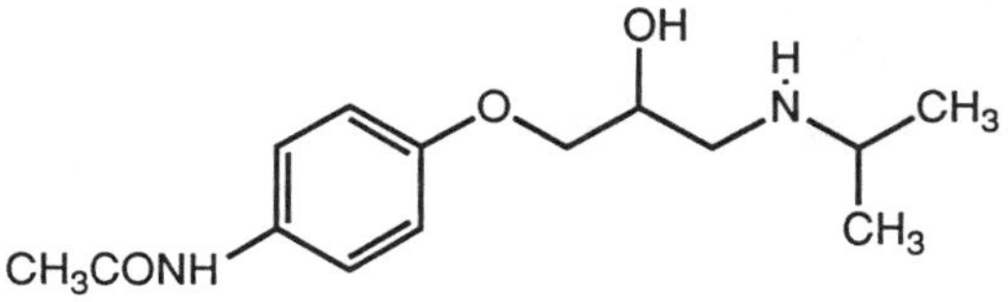

OH
H
O N CH_3
CH_3CONH CH_3

18a

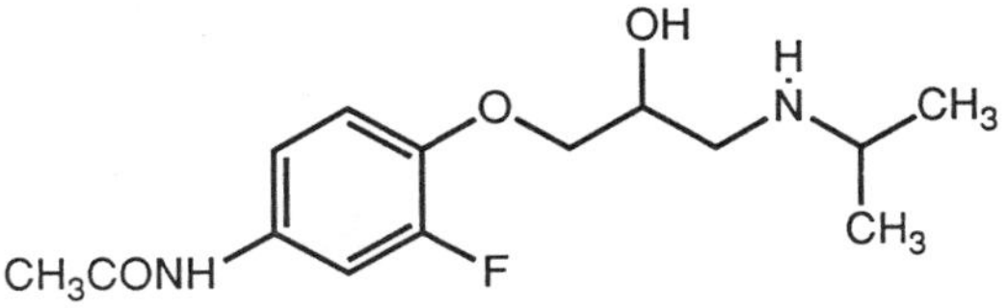

OH
H
O N CH_3
CH_3CONH F CH_3

18b

F OH
H
O N CH_3
CH_3CONH F CH_3

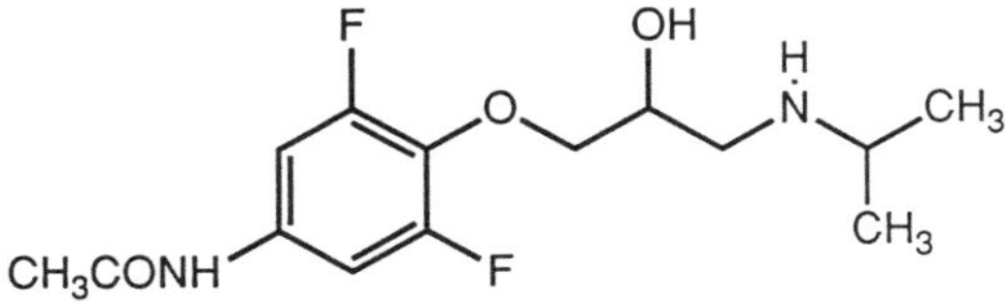

18c

Figure 14. Fluorinated analogs (**18b,c**) of the β_1-selective antagonist practolol (**18a**).

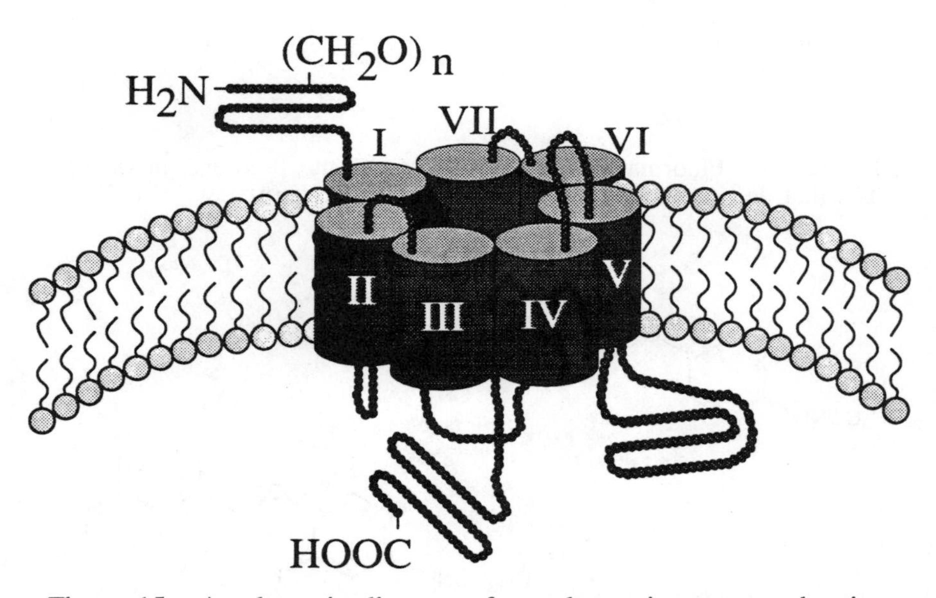

Figure 15. A schematic diagram of an adrenergic receptor showing the seven transmembrane spanning domains.

tural features of the receptor proteins that define α- and β-adrenergic selectivity. With this in mind, we have initiated a program to explore differential effects on binding to the fluorinated analogs, using α_2- and β_2-adrenergic receptors containing single amino acid substitutions.

Fluorine-Substituted Retinal and Rhodopsin. Liu and coworkers (*31*) have found that fluorine substitution on carbon-10 of retinal alters the photochemical and spectral properties of the derived rhodopsin. They proposed that a glutamate residue (Glu^{122}) on the visual protein, situated near C-13 of opsin-bound retinal, interacts with the electron-rich C-F bond. As noted above, rhodopsin and β-adrenergic receptors have structural homology. A relationship between the effects of fluorine substitution on the biochemestry of adrenergic agonists and of retinal at this time can be only a matter of speculation.

Summary

Biochemical studies of fluorinated neuroactive amines have proved to be extremely rewarding. In addition to studies of receptor mechanisms, they have had, and continue to have, applications in a multitude of other studies, including research on the mechanisms of transport, storage, release, metabolism and action of chemical neurotransmitters.

Acknowledgments

The research reviewed in this report has resulted from the diligent work of many scientists in several disciplines. The essential contributions of these numerous coworkers and collaborators are gratefully acknowleged.

Literature Cited

1. Kirk, K. L.; Cohen, L. A. *J. Am. Chem. Soc.* **1971**, *93*, 3060.
2. Kirk, K. L.; Cohen, L. A. *J. Am. Chem. Soc.* **1973**, *95*, 4619-4624.
3. Kirk, K. L.; Nagai, W.; Cohen, L. A. *J. Am. Chem. Soc.* **1973**, *95*, 8389-8392.
4. Kirk, K. L.; Cohen, L. A. in *Biochemistry Involving Carbon-Fluorine Bonds*; Filler, R., Ed.; ACS Symposium Series No. 28; American Chemical Society. Washington, D. C., 1976, pp 23-36.
5. Kirk, K. L. *J. Org. Chem.* **1976**, *41*, 2373-2376.
6. Goldberg, L. I.; Kohli, J. D.; Cantacuzene, D.; Kirk, K. L.; Creveling, C. R. *J. Pharmacol. Exp. Therapeut.* **1980**, *213*, 509-513.
7. Nimit, Y.; Cantacuzene, D.; Kirk, K. L.; Creveling, C. R.; Daly, J. W. *Life Sciences* **1980**, *27*, 1577-1585.
8. Firnau, G.; Garnett, S.; Marshall, A. M.; Seeman, P.; Tedesco, J.; Kirk, K. L. *Biochem. Pharmacol.* **1981**, *30*, 2927-2930.
9. Firnau, G.; Sood, S.; Chirakal, R.; Nahmias, C.; Garnett, E. S. *J. Neurochem.* **1987**, *48*, 1077-1082.
10. Creveling, C. R.; Kirk, K. L. *Biochem. Biophys. Res. Commun.* **1985**, *130*, 1123-1131.

11. Firnau, G.; Sood, S.; Pantel, R.; Garnett, E. S. *Mol. Pharmacol.* **1980**, *19*, 130-133.
12. Goldstein, D. S.; Chang, P. C.; Eisenhofer, G.; Miletich, R; Finn, R.; Bacher, J.; Kirk, K. L.; Bacharach, S.; and Kopin, I. J. *Circulation* in press.
13. Cantacuzene, D.; Kirk, K. L.; McCulloh, D. H.; Creveling, C. R. *Science* **1976**, *204*, 1217-1219.
14. Kirk, K. L.; Cantacuzene, D.; Nimitkitpaisan, Y.; McCulloh, D.; Padgett, W. L.; Daly, J. W.; Creveling, C. R. *J. Med. Chem.* **1979**, *22*, 1493-1497.
15. Daly, J. W.; Padgett, W.; Creveling, C. R.; Cantacuzene, D.; and Kirk, K. L. *J. Neuroscience* **1981**, *1*, 49-59.
16. Auerbach, D. A.; Klein, D. C.; Kirk, K. L.; Cantacuzene, D.; Creveling, C. R. *Biochem. Pharmacol.* **1981**, *30*, 1085-1089.
17. Chiueh, C. C.; Zukowska-Grojec, Z.; Kirk, K. L.; Kopin, I. J. *J. Pharmacol. Exp. Therapeut.* **1983**, *225*, 529-533.
18. Kirk, K. L.; Cantacuzene, D.; Collins, B.; Chen, G. T.; Creveling, C. R. *J. Med. Chem.* **1982**, *25*, 680-684.
19. Kirk, K. L.; Olubajo, O.; Buchhold, K.; Lewandowski, G. A.; Gusovsky, F.; McCulloh, D.; Daly, J. W.; Creveling, C. R. *J. Med. Chem.* **1986**, *29*, 1982-1988.
20. Adejare, A.; Gusovsky, F.; Creveling, C. R.; Daly, J. W.; Kirk, K. L. *J. Med. Chem.* **1988**, *31*, 1972-1977.
21. Chen, G.; Kirk, K. L.; manuscript in preparation.
22. Nishikawa, M.; Kanno, M.; Kuriki, H.; Sugihara, H.; Motohashi, M.; Itoh, K.; Miyashita, O.; Oka, Y.; Sanno, Y. *Life Sciences* **1975**, *16*, 305-314.
23. Thakker, D. R.; Yagi, H.; Sayer, J. M.; Kapur, U.; Levin, W.; Chang, R. L.; Wood, A. W.; Conney, A. H.; Jerina, D. M. *J. Biol. Chem.* **1984**, *259*, 11248-11256.
24. DeBernardis, J. F.; Kerkman, D. J., Winn, M.; Bush, E. N.; Arendsen, D. L.; McClellan, W. J.; Kyncl, J. J.; Basha, F. Z. *J. Med. Chem.* **1985**, *28*, 1398-1404.
25. Sugihara, H.; Ukawa, K.; Kuriki, H.; Nishikawa, M.; Sanno, Y. *Chem. Pharm. Bull.* **1977**, *25*, 2988-3002.
26. Kocjan, M.; Hodoscek, M.; Solmajer, T.; Hadzi, D. *Eur. J. Med.* **19**, *1984*, 55-59.
27. Muszkat, K. A. in *Progress in Catecholamine Research, Part A: Basic Aspects and Peripheral Mechanisms*, Dahlstrom, A.; Belmaker, R. H.; Sandler, M., Eds., Neurology and Neurobiology, Vol. 42A, Alan R. Liss, Inc., New York, 1988, pp 387-391.
28. Kaiser, C.; Jen, T.; Garvey, E.; Bowen, W. D.; Colella, D. F.; Wardell, J. R., Jr. *J. Med. Chem.* **1977**, *20*, 687-692.
29. Adejare, A.; Nie, J.-y.; Hebel, D.; Brackett, L. E.; Choi, O.; Gusovsky, F.; Padgett, W. L.; Daly, J. W.; Creveling, C. R.; Kirk, K. L., manuscript in preparation.
30. For example, Fraser, C. M.; Arakawa, S.; McCombie, W. R.; Venter, J. C. *J. Biol. Chem.* **1989**, *264*, 11754-11761.

31. Fukada, Y.; Okano, T.; Schichida, Y.; Yoshizawa, T.; Trehan, A.; Mead, D.; Denny, M.; Asato, A.; Liu, R. S. H. *Biochemistry* **1990**, *29*, 3133-3140, and references therein.

RECEIVED October 25, 1990

Chapter 10

Aldolases in Synthesis of Fluorosugars

C.-H. Wong

Department of Chemistry, Research Institute of Scripps Clinic, 10666 North Torrey Pines Road, La Jolla, CA 92037

Aldolases hold potential for convergent synthesis of fluorocarbohydrates. There have been more than 20 aldolases isolated, eight of which have been explored for organic synthesis. This presentation describes the application of fructose-1,6-diphosphate aldolase, 2-deoxyribose-5-phosphate aldolase and sialic acid aldolase to the synthesis of fluorosugars.

Fluorinated sugars are a group of compounds with potential value as pharmaceuticals or pharmacological probes (*1*). Although many nucleophilic and electrophilic fluorinating reagents have been reported for the synthesis of fluorosugars, most of the reagents are not readily available and are difficult to handle. We envisage that fluorosugars may be synthesized enzymatically via a convergent approach in which the fluorinated building blocks are prepared with inexpensive, safe and easy-to-handle fluorinating reagents such as inorganic fluoride, diethylaminosulfur trifluoride (DAST) (*2-3*), and 1-fluoropyridinium triflate (*4*). As our interest in the development of enzymes in carbohydrate synthesis (*5*), we have been active in the use of aldolases for aldol addition reactions.

There have been more than 20 aldolases isolated, eight of which have been explored for organic synthesis (*6*). Aldolases possess two interesting common features: the enzymes are specific for the donor substrate but flexible for the acceptor component, and the stereochemistry of aldol reaction is controlled by the enzyme not by the substrates. In our previous study, we have described the use of lipases, hexokinases, glycosyl transferases and rabbit muscle aldolase for the synthesis of certain fluorosugars (*7*). This review describes our recent development in aldolase-catalyzed reactions for the synthesis of fluorosugars.

Fructose-1,6-diphosphate (FDP) aldolase (EC 4.1.2.13)

The FDP aldolase from rabbit muscle (*6-8*) or *E. coli* (*9*) has been used substantially in organic synthesis. This enzyme accepts a variety of aldehydes as acceptor substrates. Figure 1 illustrates the synthesis of 6-deoxy-6-fluoro-

0097–6156/91/0456–0156$06.00/0

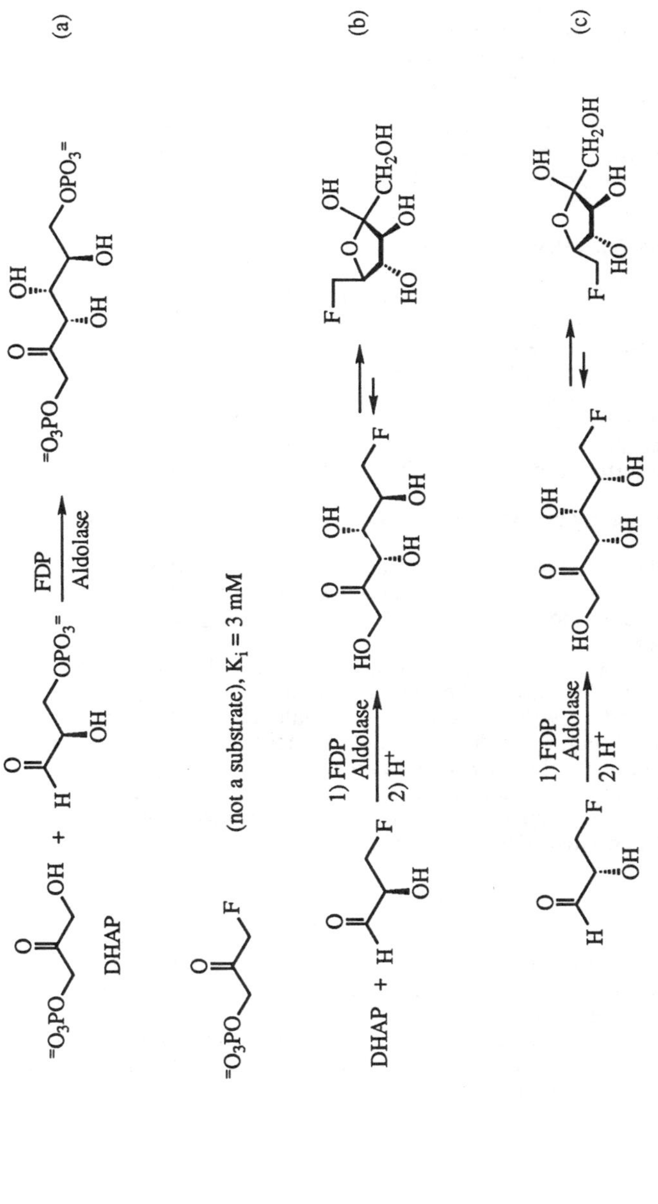

$^{=}O_3PO$ O OH + H O OH $OPO_3^{=}$ FDP Aldolase $^{=}O_3PO$ O OH OH OH $OPO_3^{=}$ (a)

DHAP

$^{=}O_3PO$ O F (not a substrate), $K_i = 3$ mM

DHAP + H O OH F 1) FDP Aldolase 2) H^+ HO O OH OH OH F ⇌ F O OH CH_2OH HO OH (b)

H O OH F 1) FDP Aldolase 2) H^+ HO O OH OH OH F ⇌ O OH CH_2OH F HO OH (c)

Figure 1. Use of rabbit muscle or *E. coli* FDP aldolase synthesis. (a) Normal reaction, (b) synthesis of 5-deoxy-5-fluoro-D-fructose, (c) synthesis of 5-deoxy-5-fluoro-L-sorbose

D-fructose and 6-deoxy-6-fluoro-L-sorbose. The fluoroaldehydes used for the aldol reactions were prepared via a nucleophilic opening of enantiomerically pure glycidaldehyde diethyl acetyl with inorganic fluoride (*9*) (Figure 2). Alternatively, both enantiomers can be prepared via a lipase-catalyzed resolution of 3-fluoro-2-acetoxypropanal diethyl acetal (*10*). The enantioselectivity of the resolution is very high, which allows for the preparation of both enantiomers with very high enantiomeric excess. One can also prepare the "R" enantiomer via an alcohol dehydrogenase-catalyzed reduction of fluoropyruvaldehyde 1,3-dithiane followed by deprotection. The alcohol dehydrogenase can be from horse liver (NAD dependent) or from *Thermoanaerobium brockii* (NADP dependent). Regeneration of the cofactor is required for a preparative synthesis. Notice that the aldol condensation strategy can be extended to the synthesis of many other fluoroketoses using other fluoroaldehyde substrates. Several fluorohexoketoses prepared with this aldolase can be converted to the corresponding aldoses catalyzed by glucose isomerase (*8*).

2-Deoxyribose-5-phosphate aldolase (DERA, EC 4.1.2.4)

The enzyme DERA from *E. coli* has recently been cloned and overexpressed in *E. coli* (*11*). It catalyzes the condensation of acetaldehyde and D-glyceraldehyde 3-phosphate to form 2-deoxyribose-5-phosphate. The enzyme also accepts different aldehydes as acceptors and donors. In addition to acetaldehyde, propionaldehyde, acetone and fluoroacetone are substrates as donors. As acceptor substrates, a broad range of aldehydes can be used (*11*). Figure 3 illustrates representative syntheses of fluorinated compounds. Work is in progress to further exploit the synthetic utility of this enzyme.

N-Acetylneuraminic Acid Aldolase (Sialic acid aldolase, EC 4.1.3.3)

Sialic acid aldolase catalyzes the condensation of pyruvate and N-acetylmannosamine to form sialic acid, an acidic sugar involved in a number of biochemical recognition processes. Like other aldolases, sialic acid aldolase accepts a number of aldoses as substrates (*12,13*). Mannose, 2-deoxyglucose, and many 6-substituted or 6-modified mannose or N-acetylmannosamine, for example, are good substrates for the enzyme. We have prepared 9-deoxy-9-fluorosialic acid and a 7,9-difluoroderivative of sialic acid using sialic acid aldolase as catalyst (Figure 4). In an attempt to prepare 3-deoxy-3-fluorosialic acid, it was found that 3-fluoropyruvate is not a substrate for the enzyme. Given the broad range of sugars which can be used as acceptor substrates for the enzyme, sialic acid aldolase appears to be a useful catalyst for the preparation of fluorinated sialic acid derivatives.

Conclusion

Enzyme catalyzed aldol condensation is a useful strategy for the convergent synthesis of fluorosugars. The fluorinated substrates can be easily prepared with readily available and easy-to-handle fluorinating reagents. With the increasing number of aldolases available, synthesis of carbohydrates and related substances based on this chemo-enzymatic strategy will experience a substantial development in the near future.

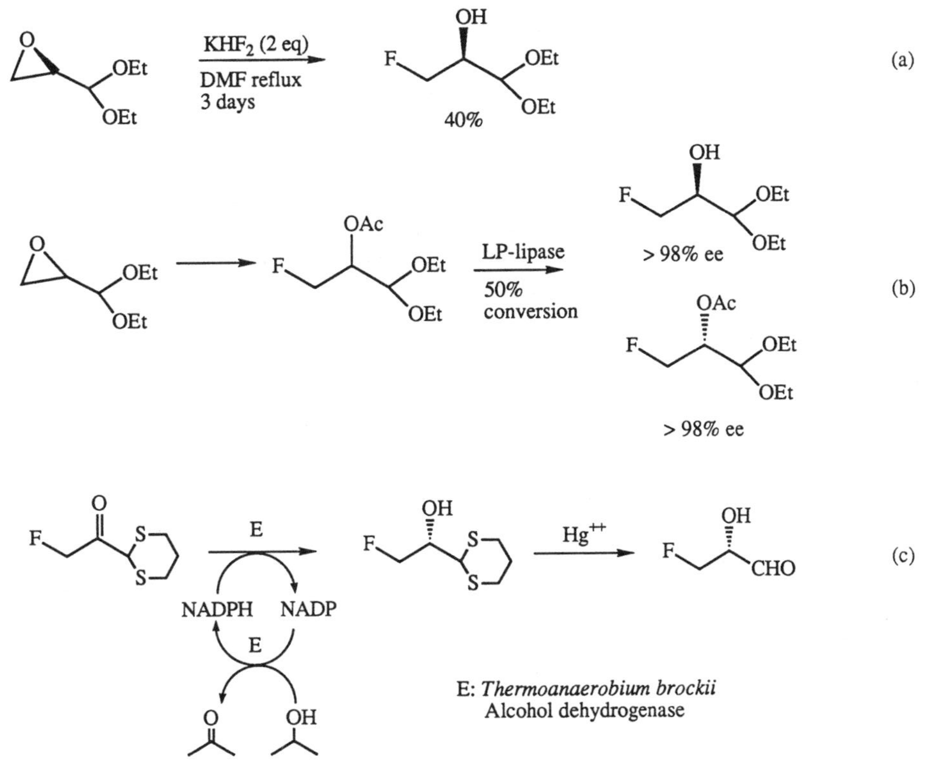

KHF_2 (2 eq)
DMF reflux
3 days
40%
(a)

LP-lipase
50%
conversion
> 98% ee
> 98% ee
(b)

E
NADPH NADP
E
Hg^{++}
E: *Thermoanaerobium brockii*
Alcohol dehydrogenase
(c)

Figure 2. Preparation of (R) and (S)-3-fluoro-2-hydroxypropanal via (a) opening of (R)-glycidal diethylacetal, (b) lipase-catalyzed resolution of a racemic precursor, (c) alcohol dehydrogenase-catalyzed asymmetric reduction

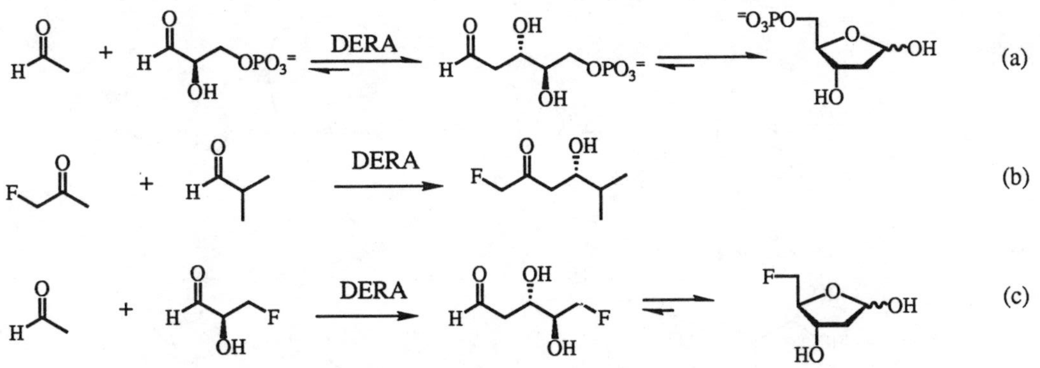

(a) H–C(=O)–CH$_3$ + H–C(=O)–CH(OH)–CH$_2$OPO$_3^{=}$ ⇌ (DERA) H–C(=O)–CH$_2$–CH(OH)–CH(OH)–CH$_2$OPO$_3^{=}$ ⇌ $^{=}O_3PO$–(furanose)–OH, HO

(b) F–CH$_2$–C(=O)–CH$_3$ + H–C(=O)–CH(CH$_3$)$_2$ → (DERA) F–CH$_2$–C(=O)–CH$_2$–CH(OH)–CH(CH$_3$)$_2$

(c) H–C(=O)–CH$_3$ + H–C(=O)–CH(OH)–CH$_2$F → (DERA) H–C(=O)–CH$_2$–CH(OH)–CH(OH)–CH$_2$F ⇌ F–(furanose)–OH, HO

Figure 3. Use of 2-deoxyribose-5-phosphate aldolase in the synthesis of fluorinated sugars. (a) Normal reaction, (b) synthesis of 1-fluoro-4-hydroxy-5-methyl-hexan-2-one, (c) synthesis of 2,5-dideoxy-5-fluororibose

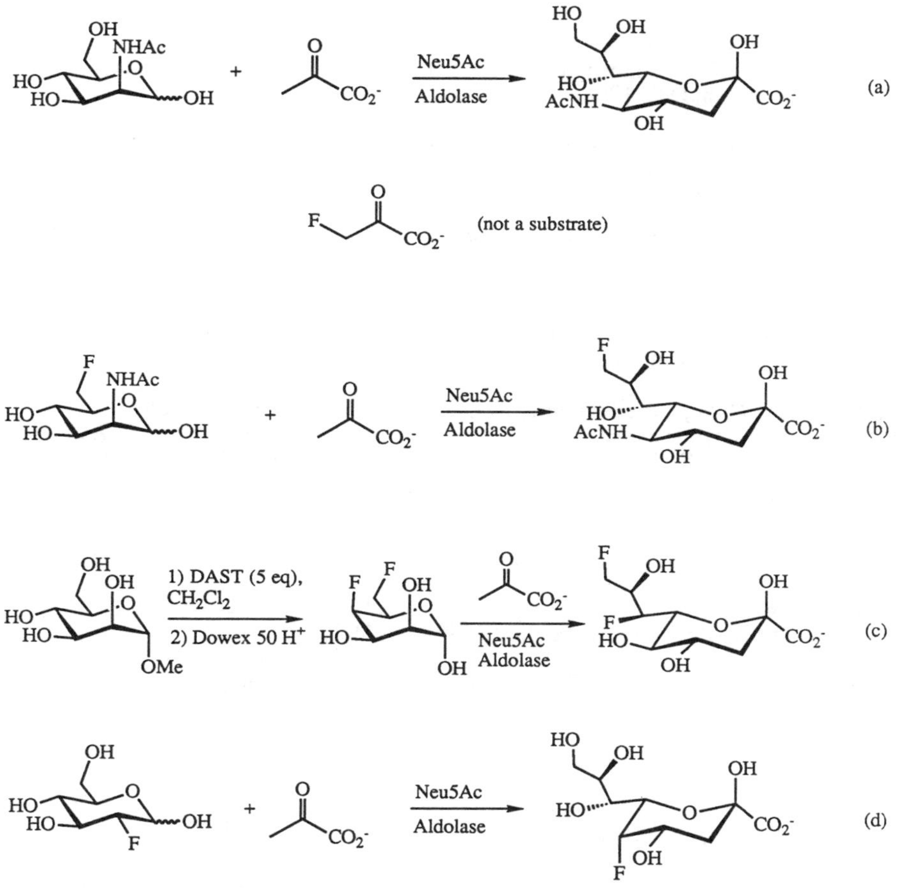

Figure 4. Sialic acid aldolase-catalyzed synthesis of fluorosialic acid and related substances. (a) Normal reaction, (b) synthesis of 9-fluoro-9-deoxy-N-acetylneuraminic acid, (c) synthesis of 7,9-difluoro-7,9-dideoxy-D-*glycero*-L-α-*altro*-2-nonulopyranosonic acid, (d) synthesis of 5-fluoro-3,5-dideoxy-D-*glycero*-α-D-*gulo*-2-nonulopyranosonic acid

Acknowledgment

I thank the contributions of many coworkers whose names are listed in the references and the compound fluoropyruvaldehyde 1,3-dithiane provided by Professor John Welch. This research was supported by the NIH GM44154-01.

Literature Cited

1. Taylor, N.F. Fluorinated carbohydrates: chemical and biochemical aspects; ACS Symposium Series No. 374; American Chemical Society: Washington, D.C. 1988.

2. Markovskij, L.N.; Pashinnik, V.E.; Kirsanov, A.V. *Synthesis*, **1973**, 787.

3. Middleton, W.J. *J. Org. Chem.* **1975**, *40*, 574.

4. Hewitt, C.D.; Silvester, M.J. *Aldrichim. Acta* **1988**, *21*, 3.

5. Wong, C.-H. *Chemtracts - organic chemistry* **1990**, *3*, 90.

6. Toone, E.J.; Simon, E.S.; Bednarski, M.D.; Whitesides, G.M. *Tetrahedron* **1989**, *45*, 5365.

7. Wong, C.-H.; Drueckhammer, D.G.; Sweers, H.M. In *Fluorinated carbohydrates: chemical and biochemical aspects*; Taylor, N.F., Ed.; ACS Symposium Series No. 374; American Chemical Society: Washington, D.C. 1988, pp 29-42.

8. Durrwachter, J.R.; Drueckhammer, D.G.; Nozaki, K.; Sweers, H.M.; Wong, C.-H. *J. Am. Chem. Soc.* **1986**, *108*, 7812.

9. von der Osten, C.H.; Sinskey, A.J.; Barbas, C.F., III; Pederson, R.L.; Wang, Y.-F.; Wong, C.-H. Ibid **1989**, *111*, 3924.

10. Pederson, R.L.; Liu, K.K.-C.; Rutan, J.F.; Chen, L.; Wong, C.-H. *J. Org. Chem.*, in press.

11. Barbas, C.F., III; Wang, Y.-F.; Wong, C.-H. *J. Am. Chem. Soc.* **1990**, *112*, 2013.

12. Auge, C.; Gautheron, C.; David, S. *Tetrahedron* **1990**, *46*, 201.

13. Kim, M.-J.; Hennen, W.J.; Sweers, H.M.; Wong, C.-H. *J. Am. Chem. Soc.* **1988**, *110*, 6481.

RECEIVED October 19, 1990

Chapter 11

Renin Inhibitors

Fluorine-Containing Transition-State Analogue Inserts

S. Thaisrivongs, D. T. Pals, and S. R. Turner

Cardiovascular Diseases Research, The Upjohn Laboratories, Kalamazoo, MI 49001

Due to its high electronegativity, fluorine effectively influences the resulting biological activity of peptidic inhibitors of the therapeutically important enzyme renin. During the enzyme-catalyzed hydrolysis of a peptidic bond of substrates by aspartyl proteases, the amide carbonyl is proposed to be hydrated to form a tetrahedral species. Chemical structures that mimic this metastable species are suggested as analogues of the transition state of the enzyme-catalyzed process and, therefore, generate compounds with very high binding affinity for those enzymes. A fluoroketone was shown to be a much more effective enzyme inhibitor than the corresponding non-fluorinated analogue. The propensity of the fluoroketone to be readily hydrated gave rise to structures that mimic the sp^3-hybridized hydrated amide carbonyl at the cleavage site of peptides. Many chemical structures that contain fluorine and have demonstrated tight binding to the enzyme renin will be summarized.

The renin-angiotensin system has been implicated in several forms of hypertension (*1*). Renin is produced mainly in the juxtaglomerular apparatus of the kidney and cleaves the circulating α-globulin angiotensinogen which is produced by the liver. The product decapeptide, angiotensin I, has no known biological activity, but it is converted to the octapeptide angiotensin II by the angiotensin converting enzyme present in the lungs. Angiotensin II is a very potent vasoconstrictor and also stimulates the release of aldosterone from the adrenal gland. This mineralocorticoid induces sodium and water retention, and this, in conjunction with vasoconstriction, can lead to an increase in blood pressure. Renin is an enzyme of very high substrate specificity, and inhibition of this specific enzyme promises to offer a highly specific therapeutic intervention in the control of hypertension (*2*). Interest in the blockade of the aspartyl enzyme renin has led to rapid development of very potent inhibitors based on the angiotensinogen sequence. The most successful approach has been

0097–6156/91/0456–0163$06.00/0

based upon the concept of a transition-state analogue of the amide bond hydrolysis(*3*). Modifications at the cleavage site to mimic the tetrahedral species have generated analogues of the minimum substrate with very high inhibitory potency in vitro (*4*).

Many peptides that contain an aldehyde function were shown to be very good inhibitors of this proteolytic enzyme (*5*). The ease with which an aldehyde forms a hemiacetal with an active site serine-OH, for example, contributes to the binding affinity since the hydrated aldehyde resembles the tetrahedral intermediate that is formed during the amide bond hydrolysis (*6*). These compounds have limited utility as an aldehyde function is metabolically unstable and can only be placed at the terminus of a peptide. Protease inhibitors that require additional binding sites at the C-terminus would be likely to lose binding affinity and also specificity. The corresponding ketone functionality can act as a pseudosubstrate of the cleavage site and peptidyl ketones are able to retain all the necessary recognition features. These compounds, however, are not likely to be the most potent inhibitors because of the very low tendency of ketones to hydrate or form hemiketals. The introduction of highly electron-withdrawing fluorine atoms adjacent to the carbonyl function, however, should render these ketones much more electrophilic therefore facilitating hydration to form tetrahedral species (*7*). Many of the known renin inhibitors that make use of the electronegativity of the fluorine atom will be summarized here.

Chemistry

A synthetic route to the protected trifluoromethyl β-amino alcohol **3** (*8*) required for the preparation of the trifluoromethyl ketone **II** is shown in Scheme 1. 2-Cyclohexylpropionic acid was deprotonated with 2.2 equivalent of lithium diisopropylamide and the resulting dianion was condensed with trifluoroacetaldehyde which was generated *in situ* from its ethyl hemiacetal. The β-hydroxy acid **1** was isolated as a racemic mixture of two diastereomers. Silylation with *tert*-butyldimethylsilyl triflate was followed by ester hydrolysis to give the acid **2**. A Curtius rearrangement with diphenylphosphoryl azide in the presence of benzyl alcohol afforded the protected β-amino alcohol **3** which was used in the preparation of the trifluoromethyl alcohol **I**. Oxidation using the Dess-Martin periodinane reagent (*9*) yielded the trifluoromethyl ketone **II** as a mixture of diastereomers. The signal for the carbonyl carbon in the ^{13}C-NMR spectrum of this ketone appeared at 94.5 ppm and this is consistent with the hydrated form of the trifluoromethyl ketone.

Synthesis of the perfluoroalkyl β-amino alcohol **5** (*10*) required for the preparation of the perfluoroalkyl ketone **VI** as shown in Scheme 2 is illustrative of the method used to prepare analogous compounds. *Tert*-butyloxycarbonyl-L-cyclohexylalaninal **4** was condensed with perfluoroethyl or perfluoropropyl lithium which was generated *in situ* by the addition of methyllithium-lithium bromide complex to the corresponding perfluoroalkyl iodide. The alcohol **5** was isolated as an epimeric mixture which was used in the preparation of peptide **IV**. Oxidation using the Dess-Martin periodinane reagent (*9*) yielded the fluoroketone **VI**.

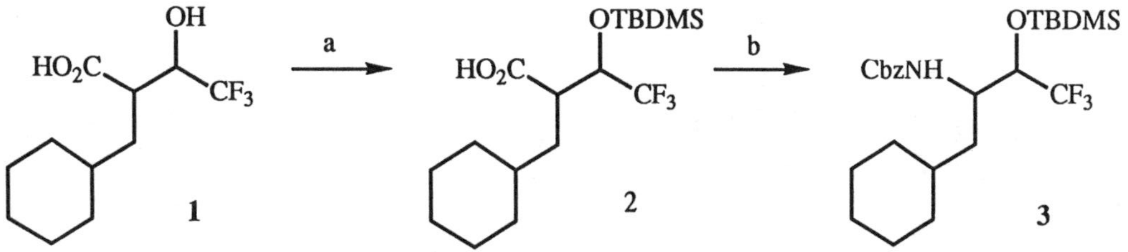

Scheme 1. Synthesis of A Trifluoromethyl Alcohol 3.
a) TBDMSOTf, Et_3N; K_2CO_3; b) DPPA, Et_3N, BnOH, hexane, reflux.

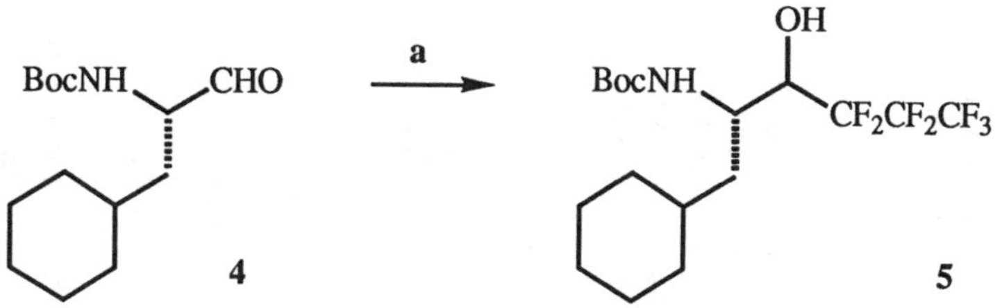

Scheme 2. Synthesis of A Perfluoroalkyl Alcohol 5.
a) $ICF_2CF_2CF_3$, $CH_3Li.LiBr$.

The straightforward synthesis of difluorostatine (*7,11*) is shown in Scheme 3. A Reformatsky reaction (*12*) of *tert*-butyloxycarbonyl-L-leucinal **4** with ethylbromodifluoroacetate in the presence of activated zinc dust afforded the diastereomeric adduct **6**. After hydrolysis of the ester, the resulting acid was used in the preparation of peptide **XI** which was separated from its epimer. Oxidation under Swern condition (*13*) of the carbinol gave the corresponding ketones as an epimeric mixture at the center which is adjacent to the carbonyl. This mixture of two compounds could be separated to give the desired isomer **XIII**.

The synthesis of 2,2-difluoro 3-amino-deoxystatine (*14*) is shown in Scheme 4. The key reaction is the stereospecific intramolecular Mitsunobu reaction (*15*). Condensation of the sodium carboxylate **7** with *p*-methoxyaniline gave the amide **8**. Intramolecular cyclization with triphenylphosphine and diethyl azodicarboxylate afforded the β-lactam **9**. Basic hydrolysis of the β-lactam ring gave the sodium salt of the protected difluoro 3-amino-deoxystatine **10**. This proved useful in the preparation of peptide **XII**.

The synthesis of the fluoroketone that combines the retroamide type bond (*16*) is shown in Scheme 5. The 2,2-difluoro-3-hydroxyester **11** from a Reformatsky reaction was converted to the primary amide **12** by treatment with ammonia in diethyl ether. Reduction of the amide with borane dimethyl sulfide and protection of the resulting amine gave the protected intermediate **13**. For the preparation of peptides **XIV** and **XV**, the hydroxy function was oxidized to the corresponding ketone using pyridinium dichromate.

The synthesis of the sulfur-containing intermediate **15** for the preparation of peptides **XVI** and **XVII** (*17*) is shown in Scheme 6. Synthetic intermediate **14** was derived from compound **6** by the reduction of the ester and ring closure after treatment with sodium hydride in dimethylforamide. The primary alcohol in compound **14** was then converted to a leaving group which was displaced by the desired thiol to give the intermediate **15**. Oxidation of the resulting sulfide then afforded the corresponding sulfone for incorporation into the desired peptide **XVI**.

Renin Inhibitory Activity

As shown in Table I, the trifluoromethyl-containing compounds with only P1-P3 (*18*) binding sites show moderate binding affinity (*8*) with micromolar IC_{50} values, which are the concentration of the inhibitors producing 50% inhibition of renin. The ketone **II** is 16 times more active than the corresponding alcohol **I**. Interestingly, binding affinity improves dramatically with small change in the chain length (*10*). The hydroxy-containing compounds **III** and **IV** with perfluoroethyl and perfluoropropyl sidechains are about 30 and 80 times respectively more active than the trifluoromethyl-containing compound **I**. As expected, the corresponding perfluoroalkyl ketones are even more active. Very impressively, the perfluoropropyl ketone **VI** is a very potent inhibitor with IC_{50} value of 3 nM. It is nearly 300 times more active than the non-fluorinated analogue **VII**.

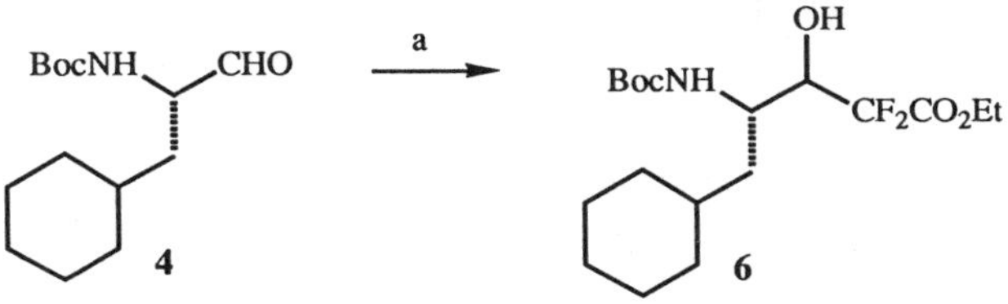

Scheme 3. Synthesis of An α,α-Difluoro-β-Hydroxyester 6.
a) BrF_2CCO_2Et, Zn, THF.

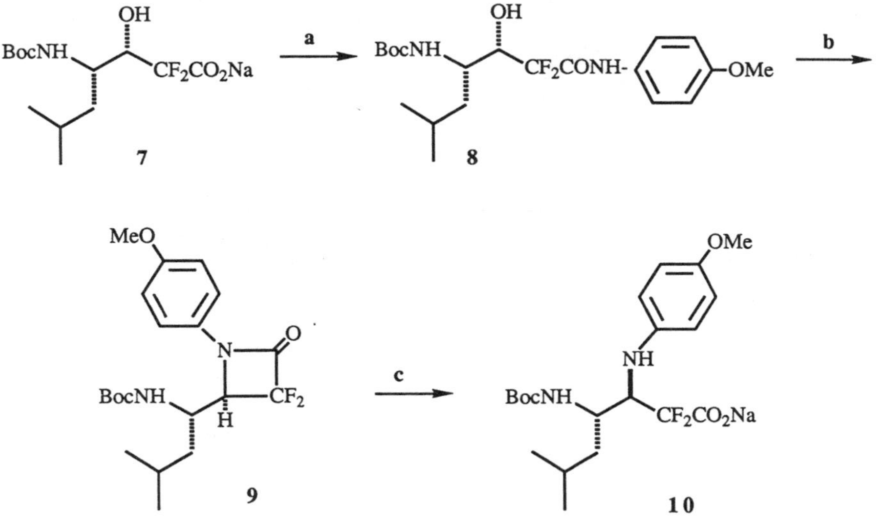

Scheme 4. Synthesis of An α,α-Difluoro-β-Aminoester 10.
a) $NH_2C_6H_4OCH_3$, BOPCl, iPr_2NEt; b) DEAD, PH_3P; c) aqNaOH.

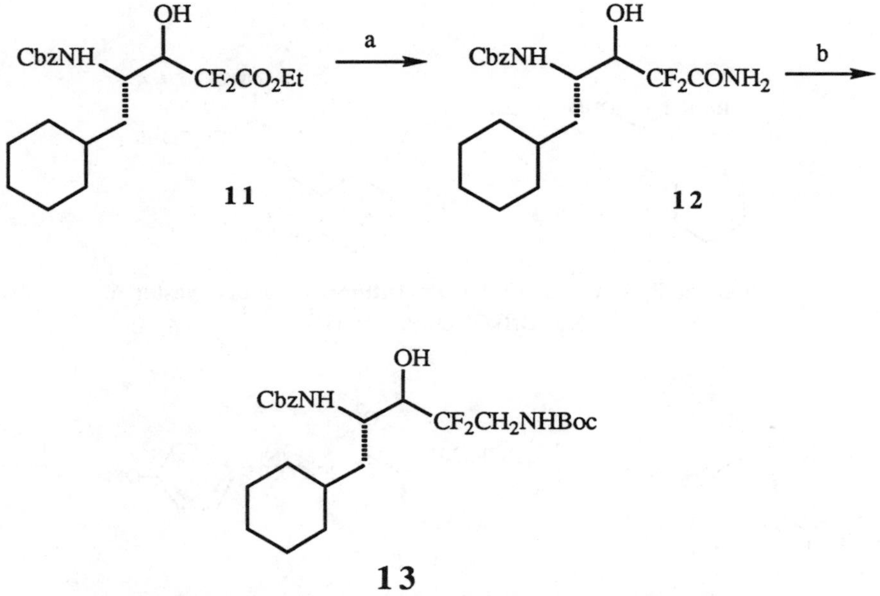

Scheme 5. Synthesis of α,α-Difluoro-Amine 13.
a) NH_3, Et_2O; b) $BH_3.Me_2S$; Boc_2O, K_2CO_3.

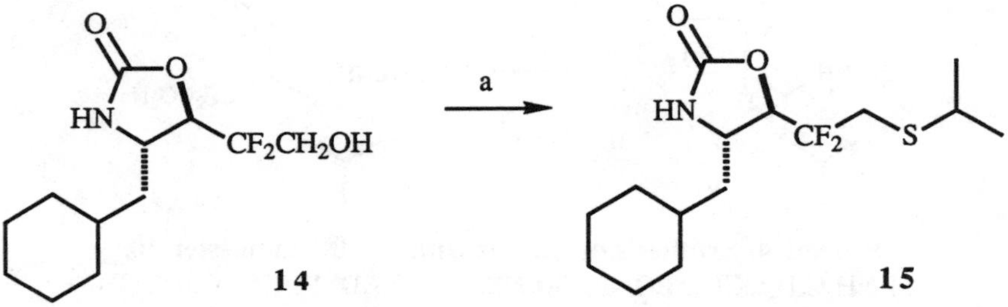

Scheme 6. Synthesis of A sulfur-containing compound 15.
a) MsCl; $HSCH(CH_3)_2$

Table I. Renin Inhibitory Activity of the Perfluoroalkyl Compounds

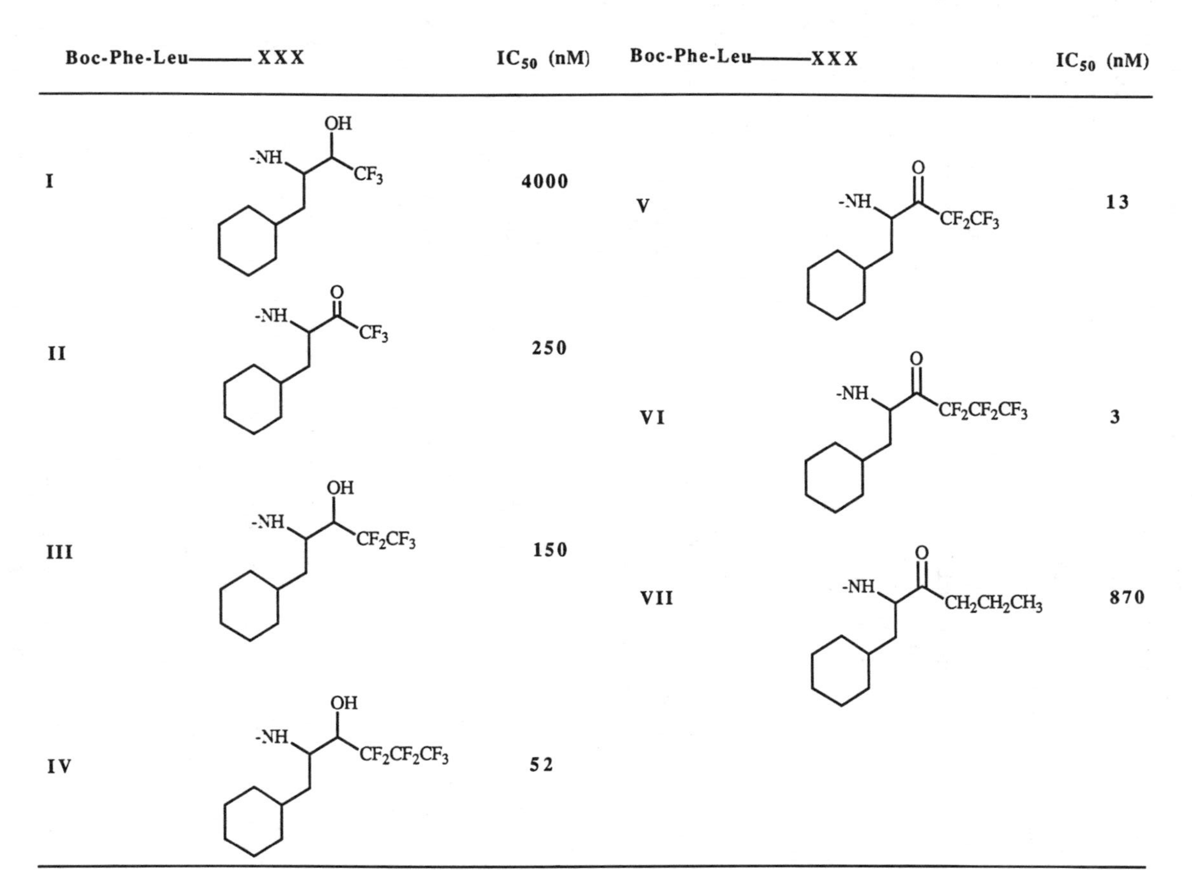

Boc-Phe-Leu——XXX	IC_{50} (nM)	Boc-Phe-Leu——XXX	IC_{50} (nM)
I	4000	V	13
II	250	VI	3
III	150	VII	870
IV	52		

Pepstatin is a naturally occurring pentapeptide that is a general aspartyl protease inhibitor (*19*). The statine residue, 4(*S*)-amino-3(*S*)-hydroxy-6-methylheptanoic acid, was proposed to act as a structural analogue of the tetrahedral species during the hydrolysis of a peptidic bond (*20*). Replacement of the leucyl valine dipeptide with statine at the cleavage site of angiotensinogen led to the preparation of a number of potent inhibitors of renin (*4*). Compound **VIII** in Table II, for example, exhibits high binding affinity with an IC_{50} value of 1.7 nM (*11*). In order to better understand the steric and electronic requirement around the statine residue, many modifications of this amino acid have been carried out. One particular analogue, 3(*S*),4(*S*)-diamino-6-methylheptanoic acid (3-aminodeoxystatine) (*21*), was predicted to be protonated on the 3-amino group at physiological pH. It was anticipated that there might be a favorable electrostatic interaction between this ammonium group and the two catalytically essential aspartic acid residues at the active site. Compound **IX**, containing the 3-aminodeoxystatine residue, inhibits renin with an IC_{50} value of 15 nM (*14*). The anticipated increase in binding affinity was not realized. It has been suggested that the potentially favorable ionic interaction is balanced by a large energy requirement for desolvation of the ammonium group as the inhibitor binds to the active site. Interest in studying the influence of the electronegative fluorine substitution on the electronic characteristics of the statine and 3-aminodeoxystatine residues led to the preparation of the corresponding α,α-difluoro-containing compounds (*11, 14, 22*) **XI** and **XII** respectively. These compounds show weaker binding affinity than the non-fluorinated congeners **VIII** and **IX**. The fluorine atoms in compound **XI** reduce the electron density on oxygen relative to the hydroxyl group of statine and thus reduce the effectiveness of the difluorostatine as a hydrogen-bond acceptor. The reduced basicity of the amino group in compound **XII** due to the adjacent electron-withdrawing fluorine atoms diminishes the effectiveness of the ammonium ion and the carboxylate ion pair. The electron-withdrawing fluorine atoms provide a beneficial effect, however, in the difluorostatone-containing compound **XIII** which exhibits very effective renin inhibition with an IC_{50} value of 0.5 nM. The carbonyl function is rendered highly electrophilic by the adjacent fluorine atoms, as the hydrated sp^3-hybridized carbon is revealed by ^{13}C-NMR spectroscopy. The corresponding non-fluorinated ketone **X** is about 70 times less active.

Other modifications of the difluoroketone as renin inhibitors are shown in Table III. The retroamide-type analogue of difluorostatone (*23*) is illustrated as in compound **XIV** which shows reasonable inhibitory potency with an IC_{50} value of 25 nM. Interestingly, the shorter congener, compound **XV**, is nearly 10 times more active. The sulfone analogue (*17*) **XVII** is also a potent inhibitor with an IC_{50} value of 1 nM. As expected, the corresponding hydroxy analogue **XVI** is not as active.

Summary

The use of the transition-state analogue concept is an effective approach in the design of potent enzyme inhibitors. For the aspartyl proteases, structures that mimic the tetrahedral hydrated amide formed during the peptidic bond hydrolysis led to the preparation of compounds with very high binding affinity to the enzymes. In the

Table II. Renin Inhibitory Activity of Compounds Containing Statine Analogues

Boc-Phe-His——XXX——Ile-Amp	XXX	IC_{50} (nM)
VIII	-NH-CH(CH₂CH(CH₃)₂)-CH(OH)-CH_2CONH-	1.7
IX	-NH-CH(CH₂CH(CH₃)₂)-CH(NH₂)-CH_2CONH-	15
X	-NH-CH(CH₂CH(CH₃)₂)-C(=O)-CH_2CONH-	34
XI	-NH-CH(CH₂CH(CH₃)₂)-CH(OH)-CF_2CONH-	12
XII	-NH-CH(CH₂CH(CH₃)₂)-CH(NH₂)-CF_2CONH-	340
XIII	-NH-CH(CH₂CH(CH₃)₂)-C(=O)-CF_2CONH-	0.5

Table III. Renin Inhibitory Activity of Compounds Containing Retroamide-type and Compounds Containing Sulfur

Compound	Structure	IC_{50} (nM)
XIV	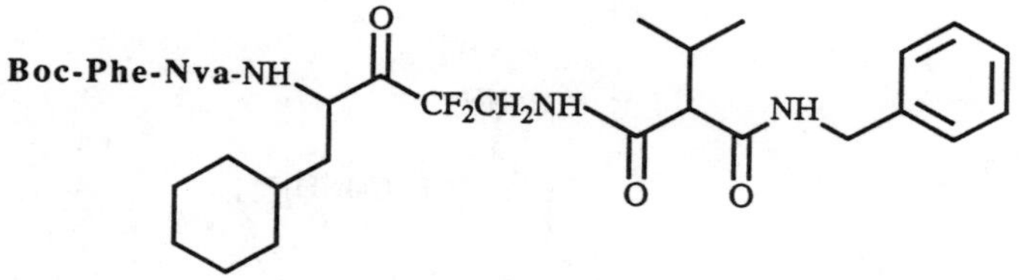 Boc-Phe-Nva-NH, O, CF_2CH_2NH, NH, O, O	25
XV	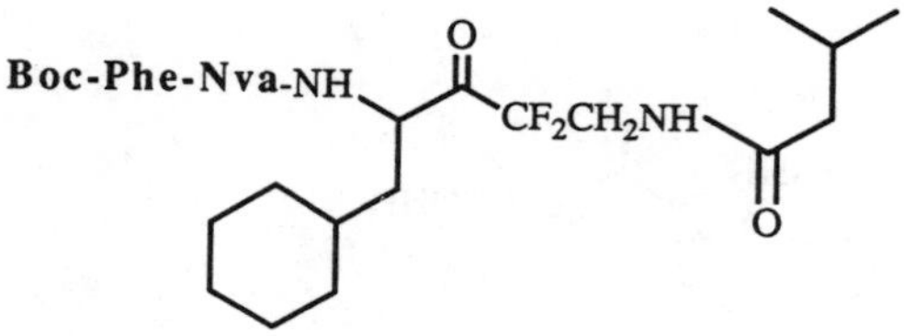 Boc-Phe-Nva-NH, O, CF_2CH_2NH, O	3.5
XVI	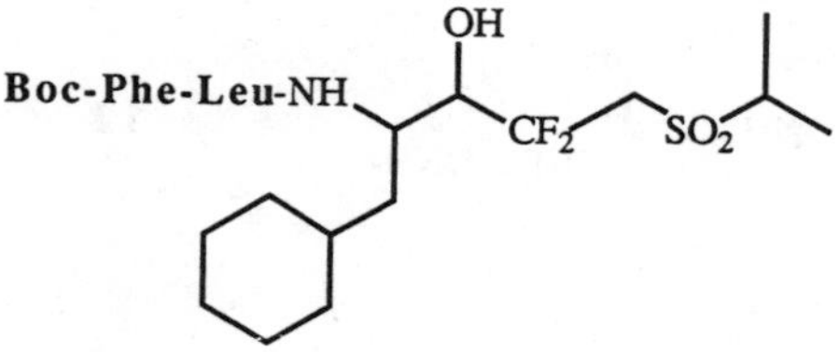 Boc-Phe-Leu-NH, OH, CF_2, SO_2	10
XVII	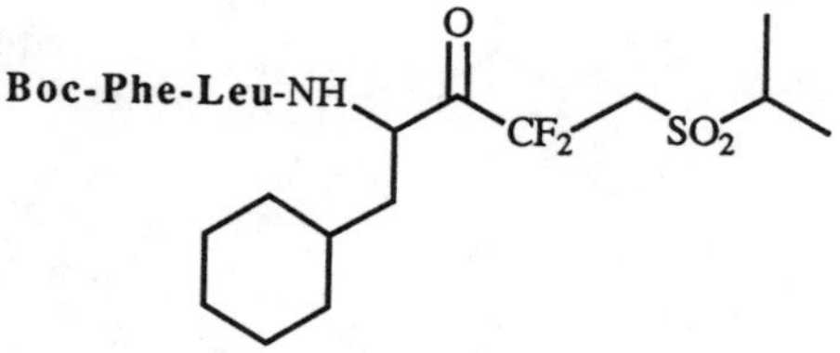 Boc-Phe-Leu-NH, O, CF_2, SO_2	1

context of fluorine-containing compounds, the hypothesis that an electron-deficient carbonyl induced by adjacent fluorination should show a high propensity to form a tetrahedral hydrated species suggested the use of fluorinated ketones as effective analogues of the transition state for amide hydrolysis. A variety of structures based on this concept led to the preparation of many very potent enzyme inhibitors illustrated in this instance for the therapeutically important enzyme renin.

Acknowledgments

We are grateful to Ms. Mae Lambert for the preparation of this manuscript.

Literature Cited

1) Davis, J.O. *Circ. Res.* **1977**, *40*, 439. Swales, J.D. *Pharmacol. Ther.* **1979**, *7*, 172.
2) Peach, M.J. *Physiol. Rev.* **1977**, *57*, 313. Ondetti, M.A.; Cushman, D.W. *Annu. Rev. Biochem.* **1982**, *51*, 283. Haber, E. *N. Engl. J. Med.* **1984**, *311*, 1631.
3) Wolfenden, R. *Transition States of Biochemical Processes*; Gandour, R.D.; Schowen, R.L.; Eds.; Plenum: New York, 1978; p 555.
4) Boger, J. *Annu. Rep. Med. Chem.* **1985**, *20*, 257. Greenlee, W.J. *Pharm. Res.* **1987**, *4*, 364.
5) Kokubu, T.; Hiwada, K.; Lato, Y.; Iwata, T.; Imamura, Y.; Matsueda, R.; Yabe, Y.; Kogen, H.; Yamazaki, M.; Iijima, Y.; Baba, Y. *Biochem. Biophys. Res. Commun.* **1984**, *118*, 929. Kokubu, T.; Hiwada, K.; Murakami, E.; Imamura, Y.; Matsueda, R.; Yabe, Y.; Koike, H.; Iijima, Y. *Hypertension, Suppl.I* **1985**, *7*, 1. Fehrentz, J.A.; Heitz, A.; Castro, B.; Cazaubon, C.; Nisato, D. *FEBS Lett.* **1984**, *167*, 273.
6) Abeles, R.H. *Drug Dev. Res.* **1987**, *10*, 221.
7) Gelb, M.H.; Svaren, J.P.; Abeles, R.H. *Biochemistry* **1985**, *24*, 1813.
8) Patel, D.V.; Rielly-Gauvin, K.; Ryono, D.E. *Tetrahedron Lett.* **1988**, *29*, 4665.
9) Dess, D.B.; Martin, J.C. *J. Org. Chem.* **1983**, *48*, 4155.
10) Sham, H.L.; Stein, H.; Rempel, C.A.; Cohen, J.; Plattner, J.J. *FEBS Lett.* **1987**, 220, 299.
11) Thaisrivongs, S.; Pals, D.T.; Kati, W.M.; Turner, S.R.; Thomasco, L.M. *J. Med. Chem.* **1985**, *28*, 1553. Thaisrivongs, S.; Pals, D.T.; Kati, W.M.; Thomasco, L.M.; Watt, W. *Ibid.* **1986**, *29*, 2080.
12) Hallinan, E.A.; Fried, J. *Tetrahedron Lett.* **1984**, 2301.
13) Omura, K.; Swern, D. *Tetrahedron* **1978**, *34*, 1651.
14) Thaisrivongs, S.; Schostarez, H.J.; Pals, D.T.; Turner, S.R. *J. Med. Chem.* **1987**, *30*, 1837.
15) Mitsunobu, O. *Synthesis* **1981**, 1.
16) Schirlin, D.; Baltzer, S.; Altenburger, J.M. *Tetrahedron Lett.* **1988**, *29*, 3687.
17) Rosenberg, S. *US Patent #4857507.*
18) Nomenclature as described by Schechter and Berger (*Biochem. Biophys. Res. Commun.* **1967**, *27*, 157); P_n-P_n' refer to the sidechain positions of the peptide substrate.

19) Umezawa, H.; Aoyagi, T.; Morishima, H.; Matsuzaki, M.; Hamada, M.; Takeuchi, T. *J. Antibiot.* **1970**, *23*, 259. Workman, R.J.; Burkitt, D.S. *Arch. Biochem. Biophys.* **1977**, *194*, 157.
20) Marciniszyn, J.; Hartsuck, J.A.; Tang, J. *J. Biol. Chem.* **1976**, *251*, 7088.
21) Jones, M.; Sueiras-Diaz, J.; Szelke, M.; Leckie, B.; Beattie, S. *Peptides, Structure and Function, Proceedings of the Ninth American Peptide Symposium*; Deber, C.M.; Hruby, V.J.; Kopple, K.D., Eds.; Pierce Chemical Co.: Rockford, IL, 1985; p 759. Arrowsmith, R.J.; Carter, K.; Dann, J.G.; Davies, D.E.; Harris, C.J.; Morton, J.A; Lister, P.; Robinson, J.A.; Williams, D.J. *J. Chem. Soc., Chem. Commun.* **1986**, 755.
22) Fearon, K.; Spaltenstein, A.; Hopkins, P.B.; Gelb, M.H. *J. Med. Chem.* **1987**, *30*, 1617.
23) Tarnus, C.; Jung, M.J.; Remy, J.-M.; Baltzer, S; Schirlin, D.G. *FEBS Lett.* **1989**, *249*, 47.

RECEIVED August 17, 1990

Chapter 12

Effect of the Fluorine Atom on Stereocontrolled Synthesis

Chemical or Microbial Methods

Tomoya Kitazume and Takashi Yamazaki

Department of Bioengineering, Tokyo Institute of Technology, O-okayama, Meguro-ku, Tokyo 152, Japan

Asymmetric transformation of fluorine-containing ketones or esters into the corresponding optically active alcohols or acids by enzymes along with the discussion on the effect of fluorinated alkyl groups during the enzymatic optical resolution or diastereoselective reactions is described.

Recent employment of optically active fluorinated compounds for biologically active substances (*1-2*) or ferroelectric liquid crystals (*3-5*) has emphasized the versatility of these chiral molecules, while few methods have been reported for the preparation of such materials in a highly diastereo- as well as enantioselective manner. On the other hand, recent investigations in this field have opened the possibility for the introduction of chirality via asymmetric reduction or optical resolution by employing biocatalysts such as baker's yeast (*6-15*) or hydrolytic enzymes (*16-20*), respectively (*21-23*), along with the conventional chemical methodology (*24-27*). Chiral materials thus obtained may also be utilized in diastereoselective reactions which create new chiral centers (*17*). In this paper, the authors would like to discuss our recent progress in the preparation of optically active fluorinated compoounds and the effect of fluorine atom(s) on the reactivity and selectivity.

Baker's Yeast-mediated Asymmetric Reduction of Fluorinated Ketones. Baker's yeast is known as one of the most convenient and readily available biocatalysts (*28*) and it promotes asymmetric inductions via the reduction, oxidation, hydrolysis, or carbon-carbon bond formation. The first functional group transformation, asymmetric reduction, is widely employed and its mode is predictable by the empirical Prelog rule (Figure 1). The reduction proceeds smoothly in many instances as shown in Table I. The absolute configurations of the newly created chiral centers, *R* in every case, would result from the *si*-face introduction of hydride. The supposition that Prelog's rule is effective in these cases led to the conclusion that all fluoromethyl moieties (R_f) used here were recognized by the enzyme as smaller than an R group. Thus, we can empirically consider that fluorine atom is only regarded as the "second smallest element" and its electronegative nature seemingly gives very little effect, if any, on the stereoselectivity. On the other hand, noteworthy is the fact that perfluoro-decanone was not a substrate. Presumably this lack of

0097–6156/91/0456–0175$06.00/0

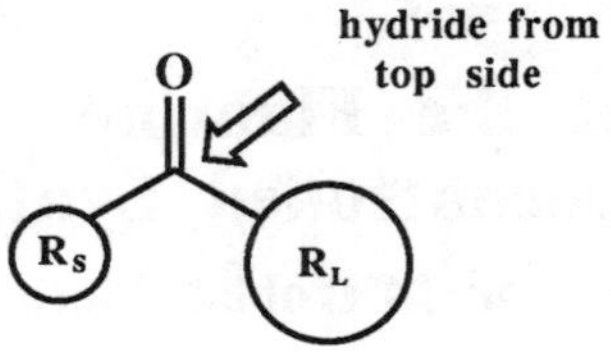

hydride from top side

O

R_S R_L

Figure 1 Prelog Rule

reactivity is caused by its poor solubility in the aqueous medium, or ready reactivity of the strongly electron deficient carbonyl group to form hydrate or complex with enzymes (*29-30*).

Table I Baker's Yeast Reduction of Fluorinated Ketones

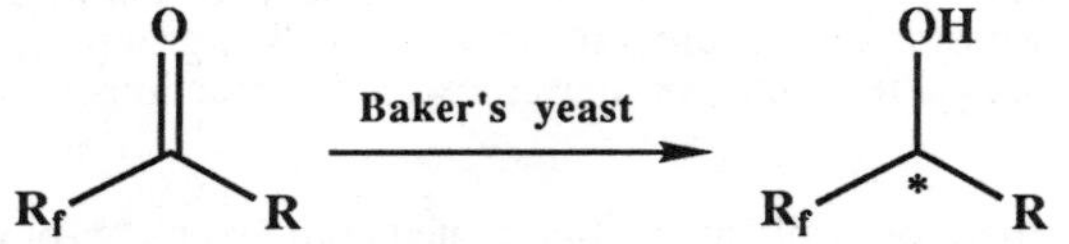

R_fC(=O)R —Baker's yeast→ R_fC*H(OH)R

R_f	R	Yield (%)	Optical Purity[a] (% ee)
CH_2F	Ph[b]	58	90 (*R*)
	$PhCH_2CH_2$	54	32 (*R*)
CHF_2	Ph	81	88 (*R*)
	CH_2CO_2Et	68	63 (*R*)
CF_3	$PhCH_2CH_2$	72	82 (*R*)
	n-C_8H_{17}	34	64 (*R*)
	CH_2CO_2Et	92	50 (*R*)
	n-C_8F_{17}		no reaction
CF_2Cl	Ph	85	7 (*R*)

[a]In the parentheses are shown their absolute configurations.
[b]This product was obtained from ethyl α-fluorobenzoylacetate via enzymatic hydrolysis and decarboxylation, which was followed by baker's yeast reduction of the resultant monofluoroacetophenone.

Lipase-catalyzed Asymmetric Hydrolysis of Fluorinated Esters. Yeast-mediated reductions form a single enantiomer predominantly and it is often difficult to find conditions which produce the opposite stereoisomer selectively. On the other hand, it is ideally possible to obtain both enantiomers in 50% yield in 100% ee via enzymatic optical resolution. Moreover, such a method could be readily optimized by use of an acyl group which facilitates stereoisomer differentiation by the employed enzyme (*31*). For example, the acetate prepared from monofluorinated α-phenethyl alcohol was transformed into the optically active secondary alcohol **(*R*)-2** in 26% ee

Table II Effect of Acyl Group and Enzyme towards Asymmetric Hydrolysis

OC(O)R / CH_2F / Ph (*rac*-1) —lipase, 40 °C→ OH / CH_2F / Ph ((*R*)-2) + OC(O)R / CH_2F / Ph ((*S*)-1)

R	Lipase[a]	Hydrolysis (%)	Optical Purity of 2 (% ee)[b]	E value[d]
CH_3	MY	34	26	1.9
i-C_3H_7	MY	33	73	9.1
n-C_4H_9	MY	46	52	4.8
n-C_7H_{15}	MY	46	25	2.0
c-C_6H_{11}	MY	11	7	1.2
Ph	MY	28	63	5.6
i-C_3H_7	MY	33	73	9.1
	O F	56	57	7.7
	NL 10	48	82	22.9
	M	30	75	9.6
	P	47	82	21.9
	PLE[c]	12	4	1.1

[a]MY (*Candida cylindracea*), OF (*Candida cylindracea*), and NL 10 (*Alcaligenes sp*. PL266) were obtained from Meito Sangyo Co., Ltd. M (*Mucor javanicus*) and P (*Pseudomonas sp.*) were purchased from Amano Pharmaceutical Co., Ltd. PLE (Pig liver esterase) was purchased from Sigma Chemical Company.
[b]Optical purity was determined by HPLC using Chiralcel OB (Daicel Chemical Industries Ltd.) [c](***R***)-**1** and (***S***)-**2** were obtained. [d]In detail, see ref 30.

when hydrolyzed with lipase MY at 34% conversion. Enhancement of optical purity to 73% ee was observed when the corresponding isobutyrate was treated with the same lipase (*33*). The best result was obtained for the transformation of the isobutyrate by lipase P, which furnished the product in 82% ee at 47% conversion. Experience has shown that one of the best combinations was the hydrolysis of acetate with lipase MY or isobutyrate with lipase P (see Table II). Immobilization (*34*) or modification (*35-36*) of these enzymes also yielded products in comparable chemical as well as optical yields.

We have also investigated the action of enzymes on phenethyl alcohols containing various fluoroalkyl groups. When the alcohols are ordered approximately increasing size of R moiety as described in Table III, a consistent trend was observed where the hydrolysis rate decreased and the optical purity of the produced alcohol increased. Another interesting point is the reversal of enantiomer recognition by lipase MY between *i*-Pr and CF_3 substituted ketones (changes of (*R*)- and (*S*)-nomenclature between CH_3 and CH_2F or CHF_2 and *i*-Pr are due to the introduction of fluorine atom(s) which alters the preference of R group relative to the phenyl substituent). This tendency suggests that interaction of fluorinated moieties with various types of

Table III Asymmetric Ester Hydrolysis from Fluoroalkylated Benzylic Alcohols

OC(O)R / R, Ph — lipase MY, 40 °C → OH / R * Ph + OC(O)R / R * Ph

rac-3 → (*R*)-4 (or (*S*)-4) + (*S*)-3 (or (*R*)-3)

R	Hydrolysis (%)	Time (h)	Optical purity of **4** (%ee)	Config.	E value
CH_3	25	1	6	(*R*)	1.2
CH_2F	34	1.5	26	(*S*)	1.9
CHF_2	35	2	30	(*S*)	2.2
i-C_3H_7	5	24	57	(*R*)	3.8
CF_3	40	24	57	(*R*)	5.2
CF_2Cl	25	7.5	73	(*R*)	8.1
CF_3CF_2	23	23	racemic	--	--
i-C_3F_7	0	24	--	--	--

electronegative functional groups, such as amines, amides, or hydroxyls surrounding the active site of enzymes, would give rise to unfavorable electronic repulsions, which in turn alter the enantiomeric recognition of enzymes. Molecules with the partially fluorinated methyl groups appear to be able to assume a favorable conformation so that the hydrogen atom confronts such electron rich functional groups. However, the unfavorable interaction which occurs when R is CF_3 or $CClF_2$ can be obviated by association to form the opposite enantiomer. Unfortunately, this concept is not generally applicable, but may be employed to rationalize this special effect of fluorination.

Table IV contains selected examples of the enzymatic resolution of esters with various structures. As was discussed above, enhancement of the optical purity was possible by changing the acyl group or the enzyme (Run 1, 2 or 6, 7). Noteworthy is the fact that *acetate 5 with a trifluoromethyl group was converted by lipase MY into the alcohol with* (*R*) *absolute configuration without exception in every case*, when the stereochemistry has been determined. However, substrates with the other fluorine-containing substituents furnished alcohols whose asymmetric configuration depended on their structures. Particularly interesting is the relationship between Runs 4 and 5 (*18*), when resolution was accompanied with simultaneous separation of diastereomers. To the best of our knowledge, this is the first example of the type of resolution of both diastereomers and enantiomers by enzymatic hydrolysis in a single transformation.

Besides secondary alcohols, asymmetric hydrolysis of α-fluoro-α-methyl-malonate to yield the corresponding half ester with (*S*)-configuration in 91% ee is also possible (*37-39*). In this compound, the chiral center generated by enzymatic hydrolysis is not epimerizable due to the absence of labile hydrogen atom at α-position to carbonyl group. Differentiation of the ester and carboxyl functionality permits the construction of compounds with either (*R*)- or (*S*)-stereochemistry (Figure 2) (*17, 40*).

Table IV Lipase-catalyzed Asymmetric Hydrolysis of Fluorinated Esters

rac-**5** (R_f–CH(OC(O)R')–R) $\xrightarrow[40\ °C]{lipase}$ *opt. active* **6** (R_f–C*H(OH)–R) + *opt. active* **5** (R_f–C*H(OC(O)R')–R)

Run	R_f	R	R'	lipase	Hydrolysis (%)	Optical purity of **6** (% ee)	
1	CH_2F	Ph	Me	MY	34	26	(*S*)
2		Ph	*i*-Pr	P	47	82	(*S*)
3		$PhCH_2CH_2$	Me	MY	34	81	(*R*)
4	CHFCl	Ph[b]	*i*-Pr	P	32	>99	(*S*)[c]
5		$PhCH_2CH_2$[b]	*i*-Pr	P	31	>99	(*S*)[c]
6	CHF_2	Ph	Me	MY	35	30	(*S*)
7		Ph	*i*-Pr	P	38	>98	(*S*)
8		$PhCH_2CH_2$	Me	P	55	73	---[a]
9		*n*-C_6H_{13}	Me	MY	51	33	(*R*)
10		CH_2CO_2Et	Me	MY	39	90	(*R*)
11	CF_3	Ph	Me	MY	40	57	(*R*)
12		$PhCH_2CH_2$	Me	MY	44	98	(*R*)
13		(*Z*)-PhCH=CH	Me	MY	28	>99	(*R*)
14		(*E*)-PhCH=CH	Me	MY	36	94	(*R*)
15		CH_2COPh	Me	MY	23	92	(*R*)
16		$CH_2COC_6H_{13}$-*n*	Me	MY	37	90	(*R*)
17		CH_2CO_2Et	Me	MY	41	96	(*R*)
18	CF_2Cl	Ph	Me	MY	25	73	(*R*)
19		$PhCH_2CH_2$	Me	MY	42	>95	---[a]
20		*n*-C_6H_{13}	Me	MY	44	88	---[a]
21	CF_3CF_2	*n*-C_8H_{17}	*i*-Pr	MY	60	50	---[a]
22		CH_2CH_2OH	Me	MY	33	99	---[a]
23	CF_3CCl_2	Ph	Me	P	27	92	---[a]

[a]Stereochemistry was unknown. [b]Diastereomeric ratios of starting material and product were as follows. 97:3 and >99:1 (R=Ph); 77:23 and >99:1 (R=$PhCH_2CH_2$).

[c]Stereochemistry at the carbon bearing hydroxy group was shown, while the configuration at the halogen-attached carbon was not clear yet.

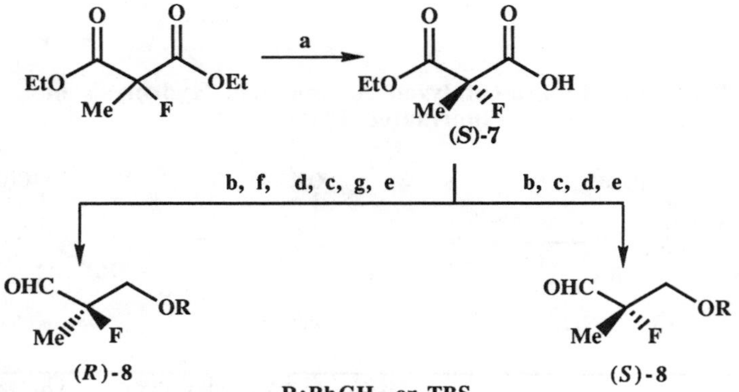

a) Lipase MY b) 1) DMF, $(COCl)_2$; 2) $NaBH_4$ c) BzlBr, NaH (or TBSCl, imidazole) d) LAH e) $(COCl)_2$, DMSO, Et_3N f) DHP, H^+ g) H^+, MeOH

Figure 2 Transformation of Optically Active Half Ester (*S*)-7 into the Corresponding Aldehydes (*R*)- or (*S*)-8

Use of Optically Active Fluorine-containing Molecules. Ethyl α,β-dihydroxy-γ,γ,γ-trifluorobutyrates, potent building blocks for the synthesis of CF_3 analogs of 6-deoxysugars, were prepared in a *syn* or *anti* selective fashion (*41*) To construct the carbon framework of these molecules effectively, two different pathways were employed.

As shown in Figure 3, the *anti*-diol was synthesized via optically active β-

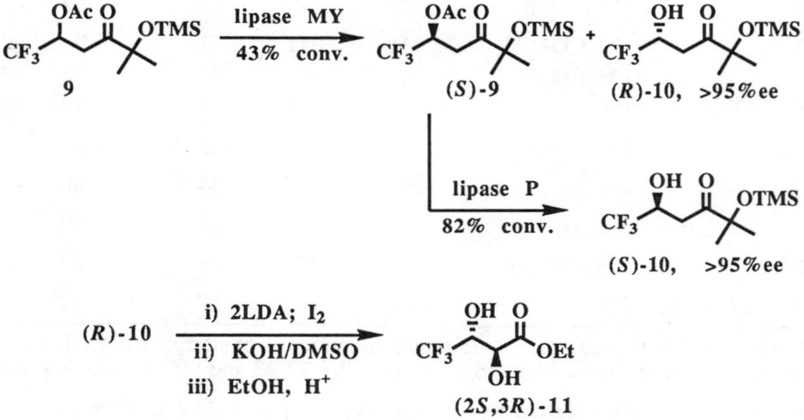

Figure 3 Preparation of Optically Active *Anti*-diol Ester (2*S*,3*R*)-11

hydroxyketone (***R***)- or (***S***)-**10**, both were resolved in >95% ee by lipase MY-catalyzed asymmetric hydrolysis. These materials were readily converted into the diol esters possessing the desired configuration via intramolecular epoxide formation and regioselective ring opening, followed by esterification (*42*).

The corresponding *syn*-isomer was, on the other hand, prepared from the potassium permanganate-mediated oxidation of trifluorocrotonate **12**, derived from the β-hydroxybutyrate. In this procedure, the diol ester with the *syn* relative configuration was the only product detected. After acetylation, ***syn*-13** was employed as a substrate for the enzymatic resolution affording both **(2*S*,3*S*)-13** and the corresponding diol ester **(2*R*,3*R*)-11** with concomitant formation of a small amount of mixture of mono-acetates ***syn*-14** (*ca* 5 to 10%, Figure 4). Two acetyl

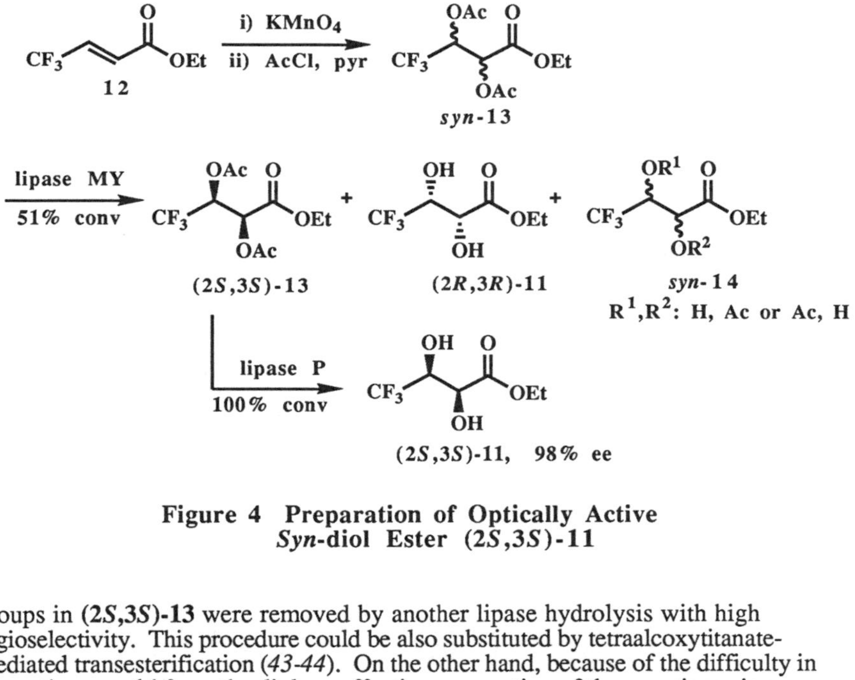

Figure 4 Preparation of Optically Active *Syn*-diol Ester (2*S*,3*S*)-11

groups in **(2*S*,3*S*)-13** were removed by another lipase hydrolysis with high regioselectivity. This procedure could be also substituted by tetraalcoxytitanate-mediated transesterification (*43-44*). On the other hand, because of the difficulty in separating ***syn*-14** from the diol, an effective preparation of the enantiomeric *syn*-isomer, **(2*R*,3*R*)-11**, has not yet been found. However, the desired conversion may be realized by searching for a better combination of enzyme or acyl group, which would yield this isomer selectively.

Diol ester **11** was next employed for the transformation of the ester functionality into an aldehyde (Figure 5). The direct conversion of ***syn*-15b** into ***syn*-17b** by reduction occurred with partial epimerization, but protected ester ***syn*-15a** afforded the corresponding alcohol ***syn*-16a** in good yield. Unfortunately, oxidation of ***syn*-16** by Swern's procedure yielded aldehydes ***syn*-17** again with epimerization in every instance. Successful conversion was eventually realized in two steps; DIBAL-H reduction of ***syn*-15a** formed the alcohol ***syn*-16a** which was then subjected to the condition of Collins' oxidation to yield ***syn*-17a** without any detectable formation of the corresponding *anti*-isomer.

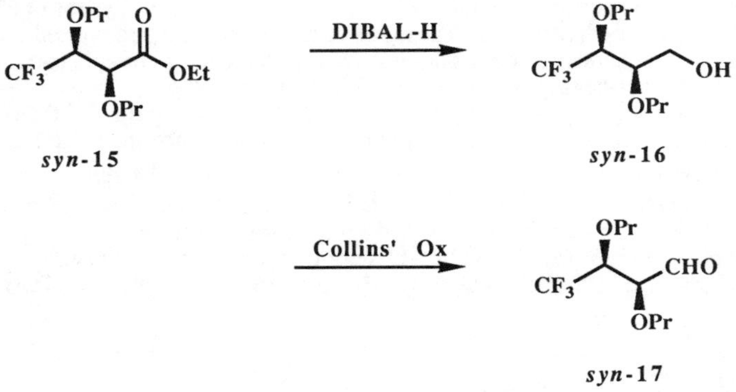

DIBAL-H

syn-15 → *syn*-16

Collins' Ox

syn-17

a:Pr=TBS, b:Pr=MOM, c:Pr_2=isopropylidene

Figure 5 Transformation of Protected *Syn*-diol Ester into the Corresponding Aldehyde

Table V Diastereoselective Cyclopropanation with Trifluoromethylated Optically Active Allylic Alcohols

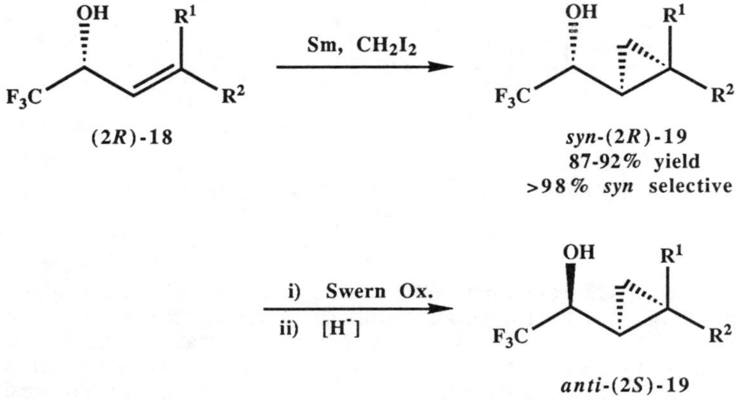

Sm, CH_2I_2

(2*R*)-18 → *syn*-(2*R*)-19
87-92% yield
>98% *syn* selective

i) Swern Ox.
ii) [H^-]

anti-(2*S*)-19

R^1	R^2	[H^-]	Selectivity[a] (*syn:anti*)
H	Ph	$NaBH_4$	32 : 68
H	Ph	L-Selectride	<1 : 99
Ph	H	$NaBH_4$	2 : 98

[a]Selectivity was checked by capillary GC (GE XE-60).

Chiral fluorinated allylic alcohols obtained by enzymatic resolution (Table IV, Run 13 and 14) were next used in diastereoselective cyclopropanation by the method reported by Molander (using the carbenoid prepared from the reaction of samarium and diiodomethane) (*45-47*). This reaction proceeded smoothly at -78 to 0 °C, but not as smoothly as with non-fluorinated substrates, when the reaction was reported to occur at -78 °C. The electron deficient nature of carbon-carbon double bond resulting from the strong electron-withdrawing effect of the trifluoromethyl group may account for this finding. While the absolute stereochemistry was not defined, the Molander's mechanism (*45-46*) predicts that the cyclopropane derivatives would be produced in a highly *syn* selective fashion (Table V). The corresponding *anti*-isomer was prepared via oxidation and reduction of the resulting carbonyl group with almost complete inversion.

These high selectivities may be explained by the transition state models described in Figure 6. The trifluoromethyl group, a steric equivalent to the isopropyl group (*48*), would be located perpendicular to the π system to minimize steric repulsion. The incoming carbenoid approaches from the side opposite to the CF_3 moiety while interacting with the hydroxyl oxygen to yield *syn*-isomer predominantly. As reported in the literature (*45-46*), a relatively large substituent such as *i*-Pr or *t*-Bu with *E*-olefin is required in this procedure for the realization of high diastereoselectivity. However, Me or *n*-Bu substituents at this position gave drastic decreases in the *syn*:*anti* ratio found (*syn*:*anti* =1:6 for Me, 1:1.4 for *n*-Bu, but 200:1 for *i*-Pr). Considering the reported selectivity, our results corroborate the earlier hypothesis that a trifluoromethyl group is steric equivalent to isopropyl moiety.

On the other hand, hydride would approach from the less hindered side as in the Felkin-Anh model (*49-50*) depicted in the next page (Figure 6) to produce cyclopropyl alcohols in an *anti* selective manner. This explanation is verified by the fact that ketone from Z-olefin (R^1=Ph, R^2=H) gave much better diastereofacial selection than the other (R^1=H, R^2=Ph) when reduced by $NaBH_4$. However,the latter ketone was transformed with high *anti* selectivity when a bulkier reducing agent, L-Selectride was employed.

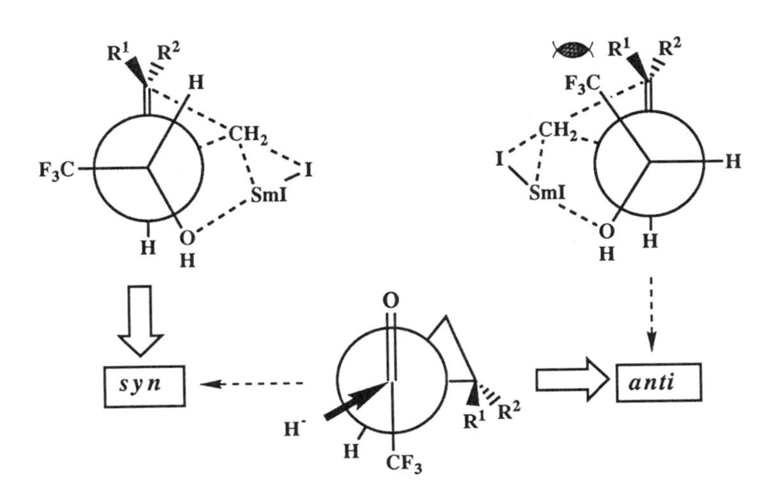

Figure 6 Possible Mechanisms for the Diastereoselective Cyclopropanation and Reduction

Conclusion. In this article, we have described our recent results on the preparation of optically active fluorinated molecules as well as the use of these compounds in diastereoselective reactions. It is our hope that this information will be helpful in attaining a deeper understanding of the nature of this "special" atom. We also look forward to the future when it will be possible to easily construct the desired optically active fluorinated molecules with the exact relative as well as absolute stereochemistry.

References

1) Welch, J. T. *Tetrahedron* **1987**, *43*, 3123.
2) *Biomedicinal Aspects of Fluorine Chemistry*; Filler, R.; Kobayashi, Y., Eds.; Kodansha & Elsevier Biomedical: Tokyo, **1982**.
3) Yoshino, K.; Ozaki, M.; Taniguchi, H.; Ito, M.; Satoh, K.; Yamasaki, N.; Kitazume, T. *Jpn. J. Appl. Phys.* **1987**, *26*, L77.
4) Johno, M.; Itoh, K.; Lee, J.; Ouchi, Y.; Takezoe, H.; Fukuda, A.; Kitazume, T. *Jpn. J. Appl. Phys.* **1990**, *29*, L107.
5) Walba, D. M.; Razavi, H. A.; Clark, N. A.; Parmar, D. S. *J. Am. Chem. Soc.* **1988**, *110*, 8686.
6) Kitazume, T.; Ishikawa, N. *Chem. Lett.* **1983**, 237.
7) Kitazume, T.; Yamazaki, T.; Ishikawa, N. *Nippon Kagaku Kaishi* **1983**, 1363.
8) Kitazume, T.; Ishikawa, N. *Chem. Lett.* **1984**, 587.
9) Kitazume, T.; Ishikawa, N. *J. Fluorine Chem.* **1985**, *29*, 431.
10) Kitazume, T.; Sato, T. *J. Fluorine Chem.* **1985**, *30*, 189.
11) Kitazume, T.; Nakayama, Y. *J. Org. Chem.* **1986**, *51*, 2795.
12) Kitazume, T.; Kobayashi, T. *Synthesis* **1987**, 187.
13) Guanti, G.; Banfi, L.; Guaragna, A.; Narisano, E. *J. Chem. Soc., Chem. Commun.* **1986**, 138.
14) Seebach, D.; Renaud, P.; Schweizer, W. B.; Züger, M. F. *Helv. Chim. Acta* **1984**, *67*, 1843.
15) Servi, S. *Synthesis* **1990**, 1.
16) Lin, J.-T.; Yamazaki, T.; Kitazume, T. *J. Org. Chem.* **1987**, *52*, 3211.
17) Yamazaki, T.; Yamamoto, T.; Kitazume, T. *J. Org. Chem.* **1989**, *54*, 83.
18) Yamazaki, T.; Ichikawa, S.; Kitazume, T. *J. Chem. Soc., Chem. Commun.* **1989**, 253.
19) Ojima, I.; Kato, K.; Nakahashi, K. *J. Org. Chem.* **1989**, *54*, 4511.
20) Keller, J. W.; Hamilton, B. J. *Tetrahedron Lett.* **1986**, *27*, 1249.
21) Jones, J. B. *Tetrahedron* **1986**, *42*, 3351.
22) Chen, C.-S.; Sih, C. J. *Angew. Chem. Int. Ed. Engl.* **1989**, *28*, 695.
23) Ohno, M.; Otsuka, M. *Org. React.* **1989**, *37*, 1.
24) Welch, J. T.; Eswarakrishnan, S. *J. Am. Chem. Soc.* **1987**, *109*, 6716.
25) Welch, J. T.; Seper, K. W. *J. Org. Chem.* **1988**, *53*, 2991.
26) Bravo, P.; Frigerio, M.; Fronza, G.; Ianni, A.; Resnati, G. *Tetrahedron* **1990**, *46*, 997.
27) Blazejewski, J. C. *J. Fluorine Chem.* **1990**, *46*, 515.
28) Crout, D. H. G.; Christen, M. In *Modern Synthetic Methods 1989*; Scheffold, R., Ed.; Springer-Verlag: Berlin, **1989**; pp 1.
29) Fearon, K.; Spaltenstein, A.; Hopkins, P. B.; Gelb, M. H. *J. Med. Chem.* **1987**, *30*, 1617.
30) Takahashi, L. H.; Radhakrishnan, R.; Rosenfield, Jr., R. E.; Meyer, Jr., E. F.; Trainor, D. A. *J. Am. Chem. Soc.* **1989**, *111*, 3368.
31) Whitesides, G. M.; Ladner, W. E. *J. Am. Chem. Soc.* **1984**, *106*, 7250.

32) Chen, C.-S.; Fujimoto, Y.; Girdaukas, G.; Sih, C. J. *J. Am. Chem. Soc.* **1982**, *104*, 7294.
33) Lin, J.-T.; Asai, M.; Ohnogi, T.; Yamazaki, T.; Kitazume, T. *Chem. Express* **1987**, *2*, 293.
34) Kitazume, T.; Okamura, N.; Ikeya, T.; Yamazaki, T. *J. Fluorine Chem.* **1988**, *39*, 107.
35) Kitazume, T.; Murata, K.; Ikeya, T. *J. Fluorine Chem.* **1986**, *32*, 233.
36) Kitazume, T.; Murata, K. *J. Fluorine Chem.* **1987**, *36*, 339.
37) Kitazume, T.; Sato, T.; Ishikawa, N. *Chem. Lett.* **1984**, 1811.
38) Kitazume, T.; Sato, T.; Kobayashi, T.; Lin, J.-T. *Nippon Kagaku Kaishi* **1985**, 2126.
39) Kitazume, T.; Sato, T.; Kobayashi, T.; Ishikawa, N. *J. Org. Chem.* **1986**, *51*, 1003.
40) Kitazume, T.; Kobayashi, T.; Yamamoto, T.; Yamazaki, T. *J. Org. Chem.* **1987**, *52*, 3218.
41) Yamazaki, T.; Okamura, N.; Kitazume, T. *Tetrahedron: Asymmetry* **1990**, in press.
42) Seebach, D.; Beck, A. K.; Renaud, P. *Angew. Chem. Int. Ed. Engl.* **1986**, *25*, 98.
43) Seebach, D.; Hungerbühler, E.; Naef, R.; Schnurrenberger, P.; Weidmann, B.; Züger, M. *Synthesis* **1982**, 138.
44) Rehwinkel, H.; Steglich, W. *Synthesis* **1982**, 826.
45) Molander, G. A.; Etter, J. B. *J. Org. Chem.* **1987**, *52*, 3942.
46) Molander, G. A.; Harring, L. S. *J. Org. Chem.* **1989**, *54*, 3525.
47) Yamazaki, T.; Lin, J.-T.; Takeda, M.; Kitazume, T. *Tetrahedron: Asymmetry* **1990**, *1*, 351.
48) Bott, G.; Field, L. D.; Sternhell, S. *J. Am. Chem. Soc.* **1980**, *102*, 5618.
49) Anh, N. T.; Eisenstein, O. *Nouv. J. Chim.* **1977**, *1*, 61.
50) Anh, N. T. *Top. Current Chem.* **1980**, *88*, 145.

RECEIVED August 27, 1990

Chapter 13

Fluoroolefin Dipeptide Isosteres

Thomas Allmendinger[1,3], Eduard Felder[2], and Ernst Hungerbuehler[1]

[1]Central Research Laboratories, [2]Research Department, Pharmaceuticals Division, Ciba–Geigy AG, CH-4002 Basel, Switzerland

Two general routes to a new class of dipeptide mimics, the fluoroolefin dipeptide isosteres, will be presented. They allow the preparation of compounds of formula **2** with a wide variety of residues R^1, R^2 and R^3 in racemic or enantiomerically pure form. Some of the mimics have been introduced into small peptides of biological interest. Some preliminary results will also be discussed.

In an attempt to overcome the major drawback of peptides being used as therapeutic agents - their rapid degradation by peptidases - the scissible peptide bond was replaced by non-hydolyzable isosteric functions (*1*, *2*). One such replacement is the trans olefin geometrically very similar to the amide bond in its most stable (transoid) conformation (*3*). Several contributions to the synthesis and application of this class of compounds **1** have been made (*4 -10*).

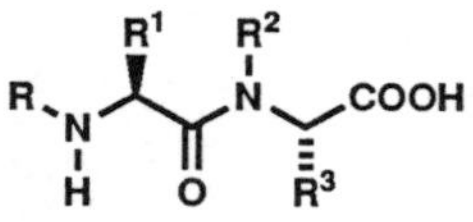

Dipeptide

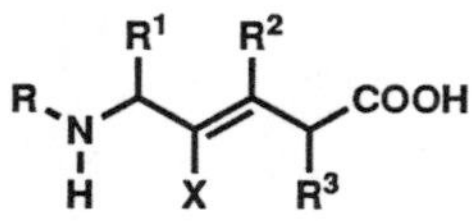

1 X = H
2 X = F

By comparison of the calculated electrostatic potentials (*11*) of trans-2-butene and 2-fluoro-2(Z)-butene with N-methyl acetamide as simple models of the peptidic bond and its isosteres (see figure 1) the fluoroolefin clearly is the better replacement of the amide bond, since it not only mimics its steric but also, at least in part, its electronic feature. Calculating dipole moments Abraham (*13*) came to similar results, but attempts to synthesize the corresponding dipeptide isostere **2** have been until now unsuccesful (*14*). As part of our ongoing program in fluoroorganic chemistry we developed two general methods for the preparation of these compounds.

[3]Current address: Development Department, Pharmaceuticals Division, Ciba–Geigy AG, CH-4002 Basel, Switzerland

0097–6156/91/0456–0186$06.00/0

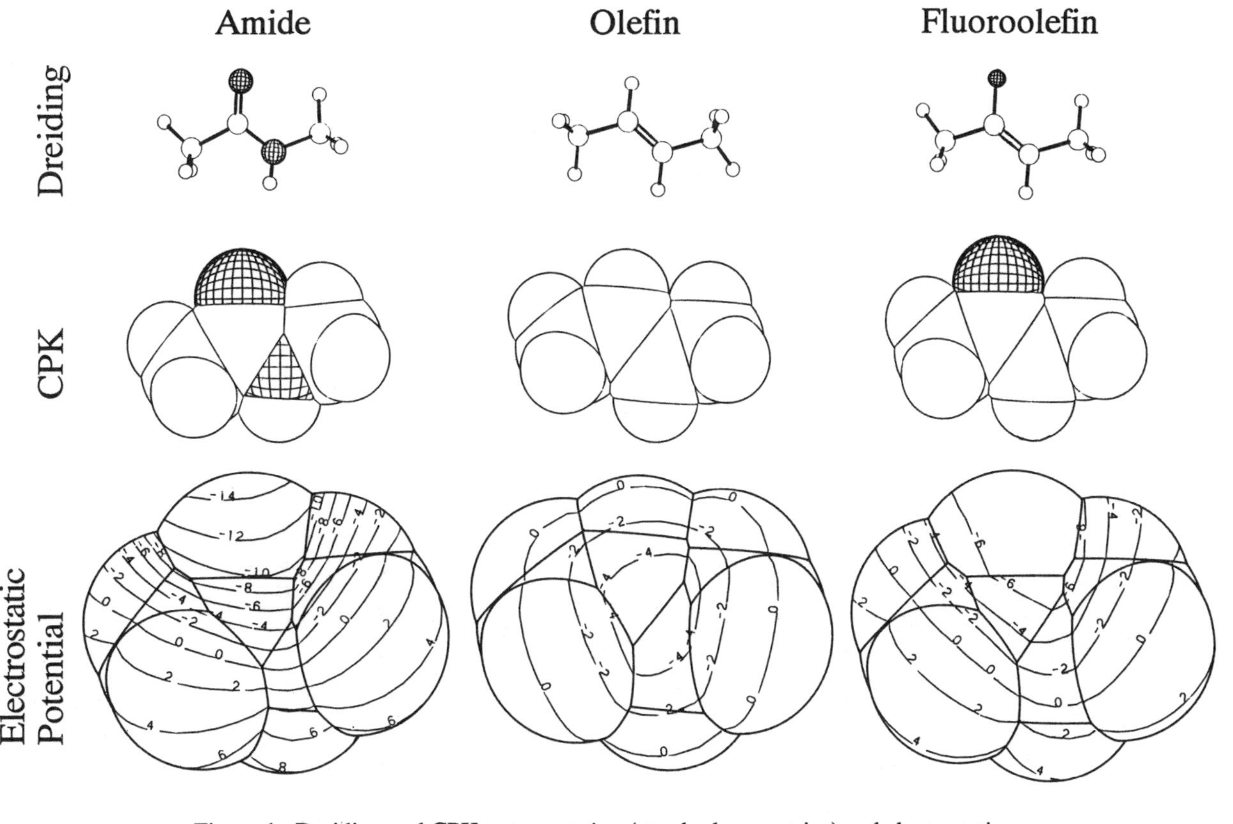

Figure 1. Dreiding and CPK representation (standard geometries) and electrostatic potential profiles of N-methyl acetamide, trans-2-butene and 2-fluoro-2(Z)-butene (*11*).

Retrosynthetic Analysis

Scheme 1 shows the key steps in a retrosynthetic manner. According to Route A the nitrogen is incorporated by reductive or alkylative amination of an aldehyde **4** via an imine. Alternatively, substitution of a primary or secondary halogen or sulfonate by a nitrogen-nucleophile proved successful as well. Route B introduces nitrogen by a formal S_N2'-type substitution of an allylic hydroxy group in hydroxyester **6**, an addition product of aldehyde **7** and ester enolates. The latter may be asymmetrically modified in both the metal M or the alcoholic part R' to induce chirality as indicated in scheme 1 (*). In the following paragraphs we describe one example of each route in detail. Since α-fluoro-α,β-unsaturated carbonyl compounds (as **4** and **7**) are important intermediates in both routes we first summarize the most general and most selective methods for their preparation (see table I).

α-Fluoro-α,β-unsaturated carbonyl compounds

According to Schlosser (*15*) the cycloaddition-reaction of chlorofluorocarbene with enol ethers and the subsequent sovolytic ring opening gives rise to specifically Z-configurated α-fluoro-α,β-unsaturated aldehydes (entry 1). The method is somewhat restricted, since electron-withdrawing substituents (e.g. R = CH_2COOMe) prevent the ring opening via carbocations (Allmendinger, T. unpublished results).

Machleidt reported the Horner-Emmons reaction of triethyl-fluoro-phosphonoacetate with aldehydes and ketones to give α-fluoro-α,β-unsaturated esters (entry 2) (*16*). In the case of aldehydes the reaction is highly E-selective (*17*), a property which was used for the preparation of fluorinated pheromones (*18*) and 5-fluoro-arachidonic acid (*19*).

Recently Ishihara (*20*) reported the Reformatsky-type reaction of methyl dichlorofluoroacetate with a number of aldehydes in the presence of acetic anhydride to give the desired esters with high selectivity (entry 3). In the course of this reaction the intermediate metallated β-hydroxy ester is acetylated and reductively eliminated in situ.

We developed a new method utilizing the previously unknown ethyl phenylsulfinyl fluoroacetate (entry 4). Its alkylation followed by thermal elimination of phenylsulfinic acid give rise to α-fluoro-α,β-unsaturated esters with high stereoselectivity, scheme 2 shows an example (Allmendinger, T. in preparation). For the first time easy accessible and stable alkyl halides (e.g. **5**, X=H,Br) instead of aldehydes (e.g. **5**, X=O) may be used for this purpose.

The PheΨ(CF=C)Pro- mimic via route A

In a HIV-directed project, inhibitors of a protease which cleaves a polyprotein-precursor between the amino acids phenylalanine and proline are desired (*21* and references therein). Therefore we designed certain peptides whose Phe-Pro unit is replaced by the isosteric fluoroolefin moiety . Scheme 3 shows the preparation for the racemic compounds.

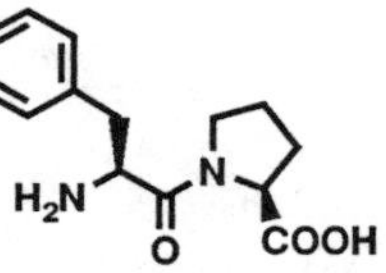

Phe-Pro

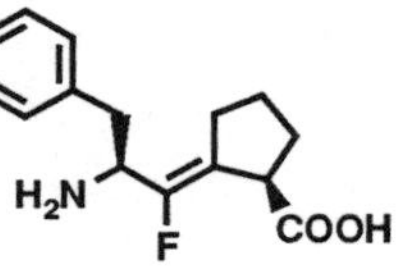

PheΨ(CF=C)Pro

Table I. α-Fluoro-α,β-unsaturated Carbonylcompounds

Entry	Reagent	Substrate	Cond.	X-CO-CF=CH-R Z/E-ratio	X
1	:CClF	EtO-CH=CH-R	a	100:0	H
2	EtOOC-CHF-PO(OEt)$_2$	OHC-R	b	<10:90	OEt
3	MeOOC-CCl$_2$F	OHC-R	c	>90:10	OMe
4	PhSO-CHF-COOEt	Br-CH$_2$-R	d	>95:5	OEt

(**a**) 1) CHCl$_2$F, 60% KOH, 18-crown-6; 2) water, sodium dodecylsulfate, reflux; (**b**) LDA, -70°C; (**c**) Zn-Cu, THF,Ac$_2$O, 50°C; (**d**) 1) base, solvent; 2) heat.

SCHEME 1

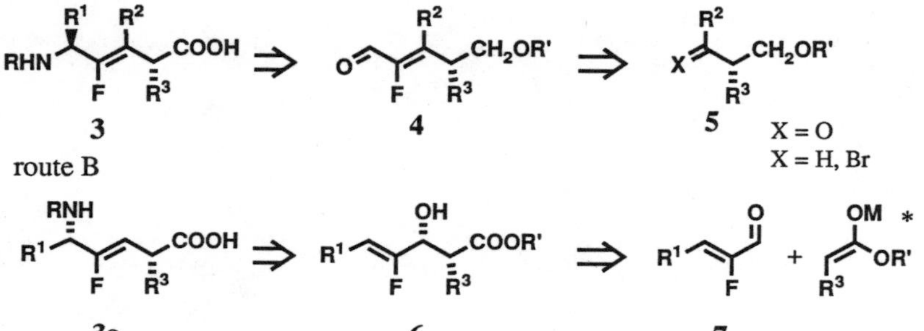

route A

3 ⇒ 4 ⇒ 5

X = O
X = H, Br

route B

3a ⇒ 6 ⇒ 7 + *

SCHEME 2

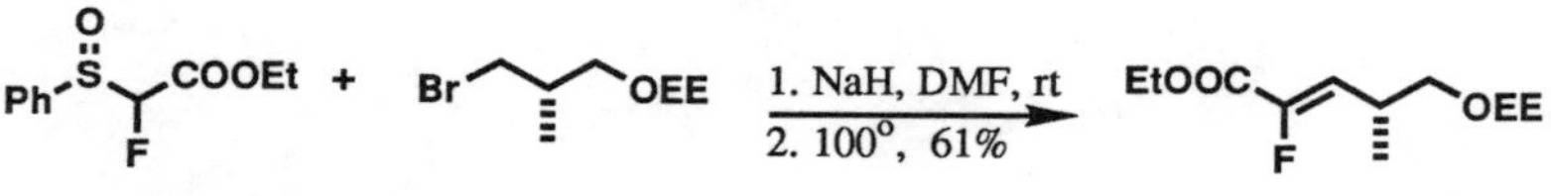

1. NaH, DMF, rt
2. 100°, 61%

EE = 1-ethoxyethyl

SCHEME 3

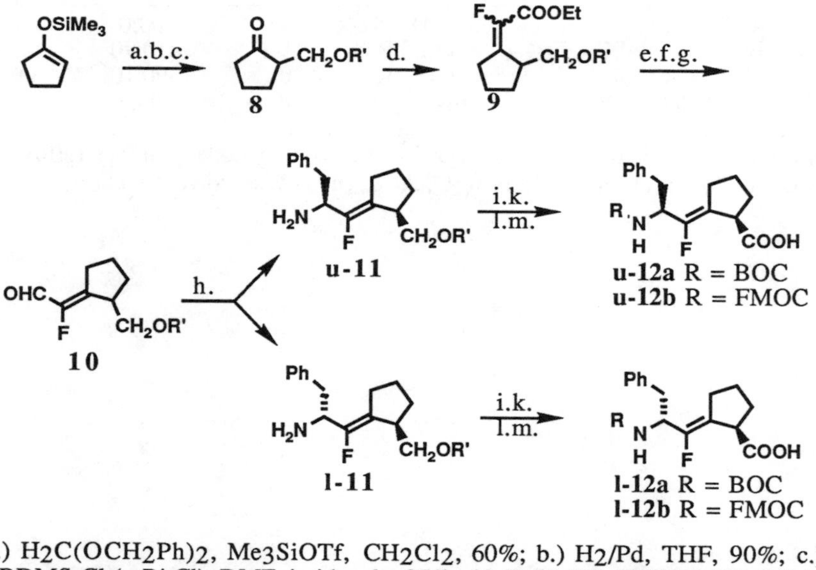

a.b.c. → 8 d. → 9 e.f.g. →

10 h. → u-11, l-11

i.k. l.m. →

u-12a R = BOC
u-12b R = FMOC

l-12a R = BOC
l-12b R = FMOC

a.) $H_2C(OCH_2Ph)_2$, Me_3SiOTf, CH_2Cl_2, 60%; b.) H_2/Pd, THF, 90%; c.) TBDMS-Cl (= R'-Cl), DMF, imidazole, 95%; d.) $(EtO)_2PO\text{-}CHF\text{-}COOEt$, LDA, THF, -20°C, 70%; e.) $H\text{-}Al(Bu^i)_2$, toluene, 80%; f.) separation, flash-chromatography; g.) Swern oxidation, 90%; h.) 1.$LiN(SiMe_3)_2$, -25°C, 2. $PhCH_2MgCl$, -70°C, 75%; i.) $(BOC)_2O$; k.) Bu_4NF, THF, 85%; l.) PDC, DMF; m.) 1. TFA, 2. FMOC-OSu, NEt_3.

Starting material was the silylprotected hydroxymethyl-cyclopentanone **8**, readily available by a Lewis-acid promoted reaction (*22*) of 1-trimethylsilyloxyclyclopentene with formaldehyde-dibenzylacetal followed by hydrogenolytic debenzylation and protection (see scheme 3).

This ketone was reacted with the fluorophosphonate (see table 1, entry 2) to give the unsaturated ester **9** as a 2:1 mixture of the E- and Z-isomers. After reduction to the E- and Z-isomeric alcohols, the latter was subjected to a Swern-oxidation giving the aldehyde **10**. By treatment with lithium hexamethyldisilazide at -25°C **10** was transformed to the corresponding silylimine, which was reacted in situ with benzylmagnesium chloride to give the amine **11** as a mixture of the l- and u-diastereomers (*23*) in a ratio of about 3:1. They have been separated by flash-chromatography and each diastereomer was treated seperately. Protection of the amine-nitrogen as its tert.-butylcarbamate, liberating of the hydroxyl group by fluoride treatment and oxidation using pyridinium dichromate in DMF gave the acids **12a** as crystalline materials. An X-ray analysis of the minor isomer confirmed the relative like-stereochemistry of this product.

This synthesis can be easily adopted for the preparation of chiral material. Corresponding intermediates have already been obtained, as outlined in scheme 4. Yeast reduction of ethyl 2-oxocyclopentanecarboxylate (*25*) afforded the optically pure hydroxyester (+)-**13**, which was further transformed to the (S)-2-silyloxymethyl cyclopentanone (-)-**8** according to scheme 4. Fluoroolefination and reduction gave optically active (-)-**14**, which can be used for the preparation of the optically pure (S,S)- and (R,S)-diastereomers of the Phe-Pro mimic.

Using similar reaction sequences we have also prepared the corresponding fluoroolefin dipeptide isosteres **15**, **16** and **17** of Phe-Phe, Gly-Gly and Phe-Gly respectively; scheme 5 shows the key intermediates. Applying the silylimine strategy to the achiral aldehyde **18**, one has only access to the Phe-Gly (or any other AA-Gly) analog in racemic form. To overcome this general limitation an alternate route was developed.

The PheΨ(CF=CH)Gly- mimic via route B

The hetero-Cope rearrangement of allylic iminoesters, established by Overman for non fluorinated molecules (*26*), leads to formal S_N2'-substitution of an hydroxy group by a nitrogen functionality.

Its application to the enantiomers of fluorinated allyl alcohols (see scheme 1) would generate optically active allylamines, a substructure of our targets. Focused on Phe-Gly isosteres **3a** (R^1 =CH_2Ph, R^3 = H) the preparation of the corresponding intermediates is shown in scheme 6.

Chlorofluorocyclopropanation of the known enol ether **19** and subsequent solvolytic ring opening gave 2-fluoro-4-phenyl-crotonaldehyde **20**. Addition of the diacetonglucose-modified titanium enolate of tert.-butylacetate (*27*) afforded the (R)-hydroxyester **21**. The (S)-hydroxyester **22** was obtained by the addition of the Li-enolate of (S)-(2-hydroxy-1,2,2-triphenylethyl) acetate [(S)-HYTRA] (*28*). Unfortunately but not surprisingly these esters do not undergo Overman-rearrangement directly via their trichloro-acetimidates because of facile β-elimination of trichloro acetamide. Therefore they were reduced to the diols **23** using lithium aluminiumhydride (see scheme 7, showing the sequence for only one enantiomer).

SCHEME 4

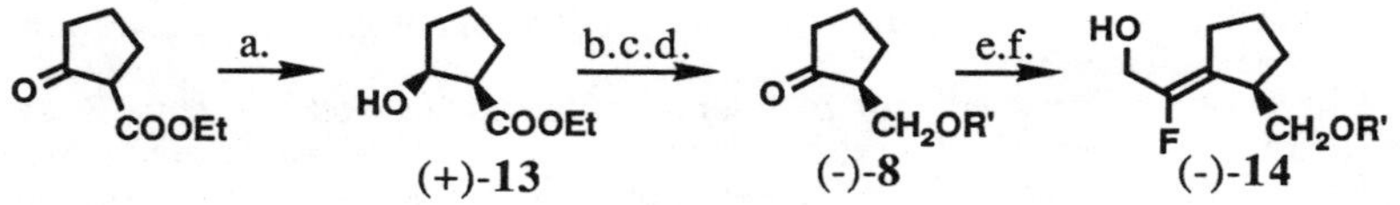

(+)-13 (-)-8 (-)-14

a.) yeast reduction, 35%; b.) $LiAlH_4$, 61%; c.) TBDMS-Cl (= R'-Cl), imidazole, DMF, 90%; d.) Swern oxidation, 60%; e.) fluoroolefination, 50%; f.) Dibal reduction, 78%.

SCHEME 5

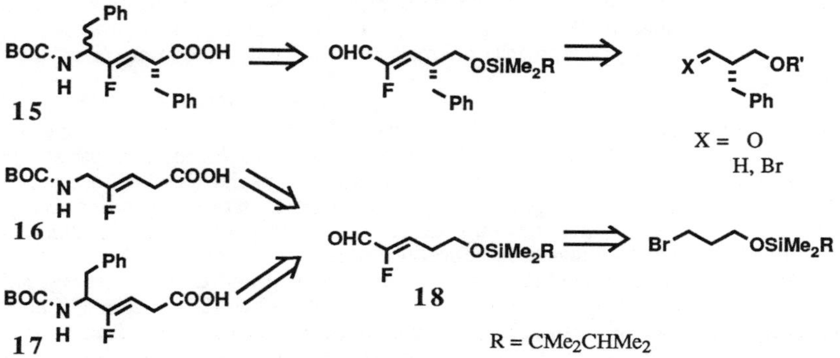

X = O
H, Br

$R = CMe_2CHMe_2$

SCHEME 6

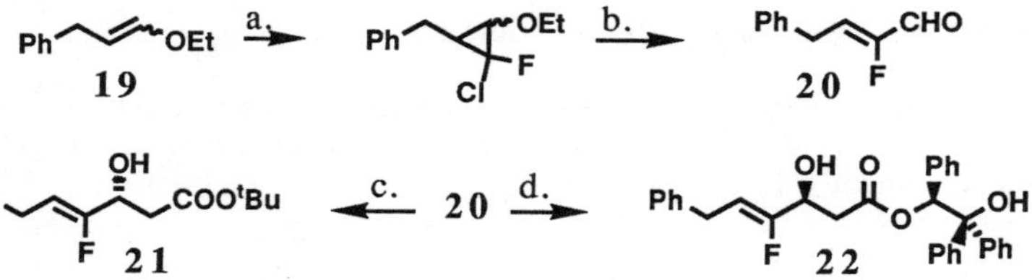

a.) CClF (see table I), 68%; b.) water, reflux, 61%; c.) $Cp(DAG\text{-}O)_2Ti\text{-}OC(OBu^t)=CH_2$, toluene, -70°C, 80%, 93% ee; d.) (S)-HYTRA, 2 LDA, -70°C, 90%, >95% ee.

With catalytic amounts of sodium hydride and two equivalents of trichloroacetonitrile, one could get the bis-iminoesters **24**, which were rearranged in refluxing xylene to **25**. The unaffected primary iminoester group was cleaved with sulfuric acid in methanol affording **26**. Replacing the N-trichloro acetyl by the BOC group and finally Jones oxidation gives the protected Phe-Gly fluoroolefin dipeptide mimics **28a** in optically pure form.

They can, as well as the aforementioned Phe-Pro-mimics **12**, be incorporated into peptides by either conventional peptide coupling or solid-phase peptide synthesis. In the latter case the FMOC-protected **28b** is necessary. The structure of related mimic-containing peptides and their biological activity is the current focus of our synthetic work.

SCHEME 7

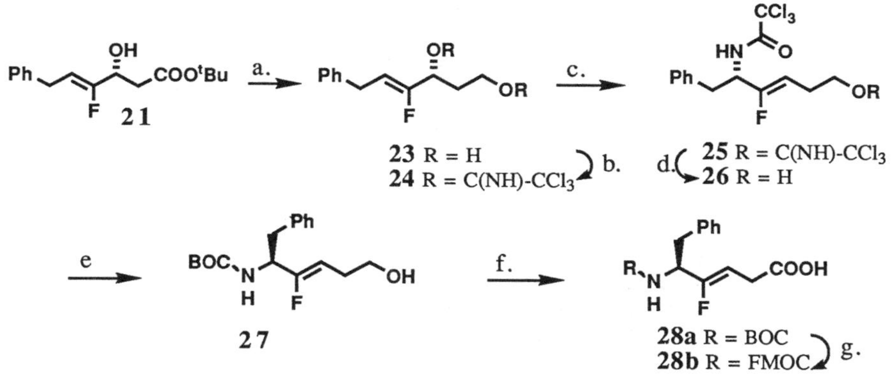

21 → a. → **23** R = H; **24** R = C(NH)-CCl_3 (b.) → c. → **25** R = C(NH)-CCl_3; **26** R = H (d.) → e → **27** → f. → **28a** R = BOC; **28b** R = FMOC (g.)

a.) LAH, ether, 0°C, 78%; b.) 1. cat. NaH, hexane, THF, 2. Cl_3CCN, ether; c.) xylene, reflux, 73%; d.) MeOH, H_2SO_4, 98%; e.) 1. NaOH, H_2O, MeOH, 2. $(BOC)_2O$, CH_2Cl_2, 81%; f.) Jones oxidation, 76%; g.) 1. TFA, 2. FMOC-OSu, NEt_3, 61%.

Peptides containing the Fluoroolefin Dipeptide Mimic and their Biological Activity

The neuropeptide Substance P attracted remarkable attention during the recent years (*5, 29-31*). We have modified its structure by replacing the Phe-Gly unit by the building block **28b** and its enantiomer using solid phase peptide synthesis (*32,33*). Using analogous conditions as for regular Fmoc amino acids, units of the type **28b** were incorporated without noticeable changes in coupling yields. The product **29** (having the same S-configuration in the replaced unit as SP itself) is nearly equipotent in a receptor binding assay compared to SP, whereas the diastereomer **30** binds 10 times weaker (table II). The order of their stimulation of guinea pig ileum is reversed, that means higher activity of the analog with "unnatural" configuration at the chiral center of the mimic. We are currently investigating this phenomenon by determining the conformation of **29** and **30** in solution, using NMR-techniques and we would like to compare their structure with the already determined structure of substance P (*34*).

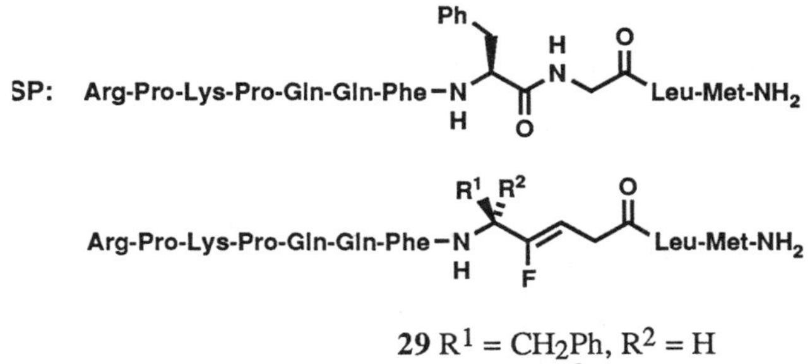

SP: Arg-Pro-Lys-Pro-Gln-Gln-Phe-NH-CH(CH_2Ph)-CO-NH-CH_2-CO-Leu-Met-NH_2

Arg-Pro-Lys-Pro-Gln-Gln-Phe-NH-C(R^1)(R^2)-CF=CH-CH_2-CO-Leu-Met-NH_2

29 $R^1 = CH_2Ph$, $R^2 = H$
30 $R^1 = H$, $R^2 = CH_2Ph$

Table II. Receptor Binding of SP and Analogues

COMPOUND	SP-BINDING IC_{50} (nm)	GP ILEUM ED_{50} (nm)
SP	1.3	0.67
29	2	15
30	20	5.9

The binding properties of **29** and **30** have not yet been compared with the corresponding non fluorinated analogues; but unlike trans alkene dipeptide isosteres, which are easily converted to compounds where the double bond is shifted into conjugation to the carbonyl group (Duplaa, H.; v. Sprecher, G.; Schilling, W., Ciba-Geigy Corp., Pharmaceuticals Division, private communication, 1989), an isomerisation was never observed with fluoroolefin dipeptide isosteres under any conditions indicating the stabilizing effect of fluorine to the double bond (*35*).

Conclusion

We have established two routes to a new class of peptide mimics, the fluoroolefin dipeptide isosteres. By appropriate selection of the precursors they allow the preparation of anaolgues of dipeptidic combinations of aminoacids bearing no other functionalities in their side chains, e.g. Gly, Ala, Val, Phe, Pro.

Acknowledgment: We thank Gisela Geiger, Andrea Zingg, Guenther Bartsch and Hans Ofner for their skillful assistance in the laboratory, Dr. Kathleen Hauser for the biology, Dr. Pascal Furet for molecular modelling and Dres. Hans Greuter, Robert W. Lang and Walter Schilling for helpful discussions.

Literature cited

1. Spatola, A. in *Chemistry and Biochemistry of Amino Acids, Peptides and Proteins*; Weinstein, B , Ed.; Marcel Dekker: New York, ,1983, Vol. 7; pp. 267-357.
2. Tourwe, D.*Janssen Chimica Acta* **1985**, *3,* 3.
3. Hann, M.M.; Sammes, P.G.; Kennewell, P.D.; Taylor, J.B. *J.Chem.Soc., Chem.Comm.* **1980,** 234; *Perkin Trans.* I **1982,** 307.
4. Cox, M.T.; Heaton, D.W.; Horbury, J *J.Chem.Soc., Chem. Comm.* **1980**, 799.
5. Cox, M.T.; Gormley, J.J.; Hayward, C.F.; Petter, N.N. *ibid* **1980**, 800.
6. Miles, N.J.; Sammes, P.G.; Kennewell, P.D.; Westwood, R. *J. Chem. Soc. Perkin Trans. I* **1985**, 2299.
7. Spaltenstein, A.; Carpino, P.A.; Miyake, F.; Hopkins, P.B. *J. Org. Chem.* **1987**, *52*, 3759.
8. Shue, Y.-K.; Carrera, Jr, G.M.; Nadzan, A.M. *Tetrahedron Lett.* **1987**, 3225.
9. Shue, Y.-K.; Tufano, M.D.; Nadzan, A.M. *ibid* **1988**, 4041.
10. Whitesell, J.K.; Lawrence, R.M. *Chirality* **1989**, *1*, 89.
11. The molecular electrostatic potentials were calculated in the monopoleapproximation using atomic charges derived from a Mulliken population analysis on MNDO wave functions. The potentials are displayed as isovalue contour lines on expanded Van der Waals surfaces of the molecules (*12*) (extra atomic radius of 1.5 A). Negative and positive values correspond to regions of attraction and repulsion respectively for a unitary positive charge.

12. Cohen, N.C. *ACS Symp. Ser.* **1979**, *112*, 371-381.
13. Abraham, R.J.; Ellison, S.L.R.; Schonholzer, P.; Thomas, W.A. *Tetrahedron* **1986**, *42*, 2101.
14. Ellison, S.L.R. *PhD. Thesis*, University of Liverpool, UK, 1984.
15. Bessiere, Y.; Savary, D.N.-H.; Schlosser, M. *Helv. Chim. Acta* **1977**, *60*, 1739.
16. Machleidt, H.; Wessendorf, R. *J. Liebigs Ann. Chem.* **1964**, *674*, 1.
17. Etemad-Moghadam, G.; Seyden-Penne, J. *Bull. Soc. Chim.France* **1985**, 448.
18. Camps, F.; Coll, J.; Fabrias, G.; Guerrero, A. *Tetrahedron* **1984**, *40*, 2871.
19. Taguchi, T.; Takigawa, T.; Igarashi, A.; Kobayashi, Y.; Tanaka, Y.; Jubiz, W.; Briggs, R.G. *Chem. Pharm. Bull.* **1987**,*35*, 1666.
20. Ishihara, T.; Kuroboshi, M. *Chem. Lett.* **1987**, 1145.
21. Dreyer, G.B.; Metcalf, B.W.; Tomaszek, Jr., T.A.; Carr, T.J.; Chandler, III, A.C.; Hyland, L.; Fakhoury, S.A.; Magaard, V.W.; Moore, M.L.; Strickler, J.E.; Debouck, C.; Meek, T.D. *Proc. Natl. Acad. Sci. USA* **1989**, *86*, 9752.
22. Murata, S.; Suzuki, M. Noyori, R. *Tetrahedron Lett.* **1980**, 2527.
23. l (like) and u (unlike) according to the nomenclature of Seebach and Prelog (*24*) are equal to a RR, (SS) and RS, (SR) relationship respectively of two asymmetric centers.
24. Prelog, V.; Seebach, D. Angew. Chem. **1982**, *94*, 696, *Angew. Chem. Internat. Edit. Engl.* **1982**, *21*, 654.
25. Deol, B.S.; Ridley, D.D. ; Simpson, G.W. *Austr. J. Chem.* **1976**, *29*, 2459.
26. Overman, L.E. *J. Am. Chem. Soc.* **1976**,*98*, 2901.
27. Duthaler, R.O.; Herold, P.; Lottenbach, W.; Oertle, K.; Riediker, M. *Angew. Chem., Int. Ed. Engl.* **1989**, *28*, 495.
28. Devant, R.; Mahler, U.; Braun, M. *Chem. Ber.* **1988**, *121*, 397.
29. *Substance P in the Nervous System* ; Porter, R.; O'Connor, M. Eds.; Ciba Foundation Symposium 91; Pitman: London, Great Britain, 1982.
30. *Proceedings of the 17th European Peptide Symposium*, Aug. 29-Sept. 3., 1982, Prague, Czechoslovakia, pp. 511-549;
31. Bartho, L.; Holzer, P. *Neuroscience* **1985**, *16*, 1.
32. Atherton, E.; Fox, H.; Harkiss, D.; Logan, L.J.; Sheppard, R. L.; Williams, B.J. *J. Chem. Soc. Chem. Commun.* **1978**, 537.
33. Rink, H. *Tetrahedron Lett.* **1987**, 3787.
34. Chassaing, G.; Convert, O.; Lavielle, S. *Eur. J. Biochemistry* **1986**, *154*, 77.
35. Dolbier, Jr., W.R.; Medinger, K.S.; Greenberg, A.; Liebmann, J.F. *Tetrahedron* **1982**, *38*, 2415.

RECEIVED August 17, 1990

Chapter 14

The Influence of Fluoro Substituents on the Reactivity of Carboxylic Acids, Amides, and Peptides in Enzyme-Catalyzed Reactions

James K. Coward[1], John J. McGuire[2], and John Galivan[3]

[1]Departments of Medicinal Chemistry and Chemistry, University of Michigan, Ann Arbor, MI 48109
[2]Department of Experimental Therapeutics, Roswell Park Memorial Institute, Buffalo, NY 14263
[3]Wadsworth Center for Laboratories and Research, New York State Department of Health, Albany, NY 12201

Two fluorinated analogs of glutamic acid have been studied for their ability to affect the biosynthesis and hydrolysis of folyl- and antifolylpoly-γ-glutamates. Polyglutamates are the predominant intracellular forms involved in folate-mediated one-carbon biochemistry and in the action of clinically useful anticancer agents such as methotrexate. Placement of fluorine atoms at either the 3- or 4- position of glutamate leads to dramatic changes in the rates of the enzyme-catalyzed biosynthesis and hydrolysis of these polyglutamates. Studies on the biosynthetic reaction, catalyzed by folypoly-γ-glutamate synthetase, revealed that DL-*threo*-4-fluoroglutamate is a chain terminating substrate whereas DL-3,3-difluoroglutamate stimulates chain elongation. Synthesis of fluorine-containing oligo-γ-glutamates and investigation of their hydrolysis, catalyzed by γ-glutamyl hydrolase, revealed that the fluorine-containing glutamyl peptides are hydrolyzed very slowly when fluorine is α- or β- to the scissile peptide bond.

The vitamin folic acid is, like most other vitamins, an artifact of isolation. In contrast to the fully oxidized pteridine coupled to p-aminobenzoylglutamate found in folic acid, intracellular folates are predominantly γ-glutamyl peptide "conjugates" containing reduced and substituted pteridine moieties (Scheme I). The biochemistry of reduced and substituted pteroyl monoglutamates has been studied extensively (*1*). However, only recently have similar detailed studies on the biochemistry and pharmacology of folate polyglutamates been carried out (*2,3*). These studies have led to the hypothesis that intracellular folylpoly-γ-glutamate flux involves a balance between synthesis, catalyzed by the enzyme folylpoly-γ-glutamate synthetase (FPGS, EC 6.3.2.17), and hydrolysis, catalyzed by γ-glutamyl hydrolase (γ-GH, EC 3.4.22.12). The biochemistry of this flux is depicted in Scheme I.

Our interest in the effect of fluorine substitution on polyglutamate metabolism arose from the observation that the DL-*threo* diastereomer of 4-fluoroglutamate

0097–6156/91/0456–0196$06.00/0

Scheme I

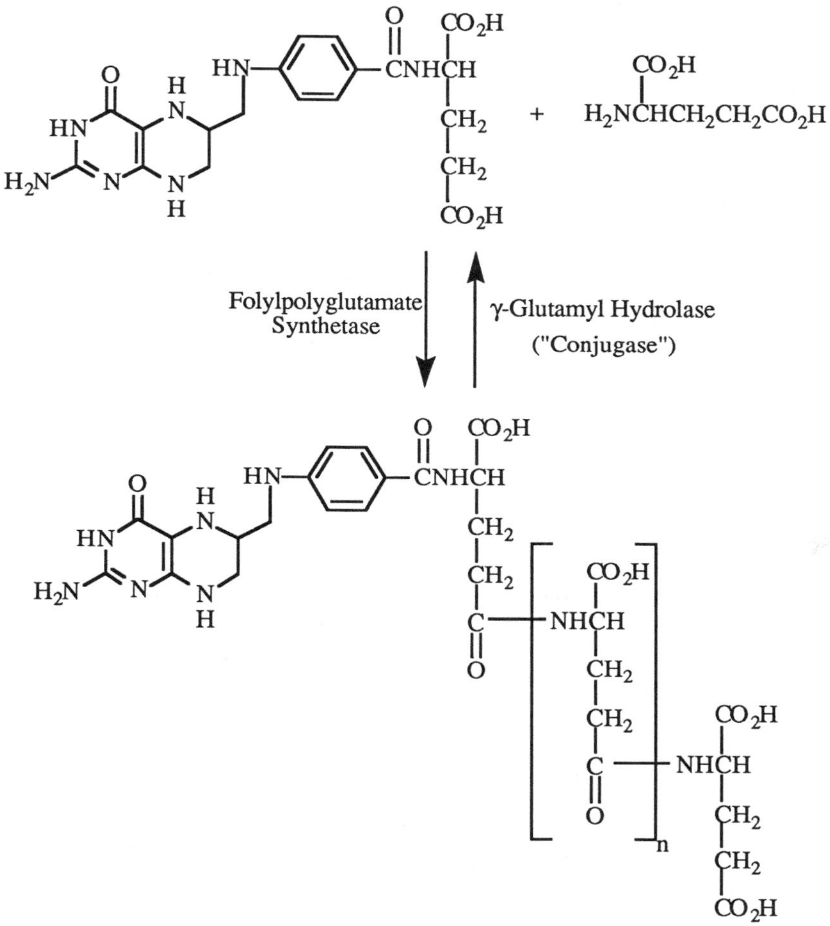

Folylpolyglutamate Synthetase

γ-Glutamyl Hydrolase ("Conjugase")

(2*S*,4*S*;2*R*,4*R*) acts as chain-terminating substrate of FPGS (*4*). In contrast, the DL-*erythro* diastereomer (2S,4R;2R,4S) is a very poor FPGS substrate but when incorporated, it too acts as a chain terminator. These results appeared to be consistent with our γ-glutamyl phosphate mechanism for the FPGS-catalyzed reaction (*5*). Thus, the conjugate base of 4-fluoroglutamate (FGlu) would be expected to be less nucleophilic than glutamate (Glu) (pK_a=ca. 2.5 vs. ca. 4.65, FGlu vs. Glu, respectively) and less likely to form the acyl phosphate intermediate. In this paper, we describe some more recent studies which extend the initial observation on FGlu and describe initial studies with another fluorine-substituted glutamate analog, DL-3,3-difluoroglutamic acid (F_2Glu).

Effects of Fluorine Substituents on Methotrexate Cytotoxicity

Synthesis and Properties of γ-Fluoromethotrexate. Following the discovery that the DL-*threo* isomer of FGlu acts as a chain terminating substrate in the FPGS-catalyzed reaction, we synthesized an analog of the anticancer drug, methotrexate (MTX, 4-amino-10-methylpteroyl-L-glutamic acid), in which Glu is replaced by FGlu (*6*). Our initial synthesis led to a mixture of all four isomers of the final product, γ-fluoromethotrexate [**1**, FMTX, 4-amino-10-methylpteroyl(DL-*erythro, threo*-4-fluoro)glutamic acid].

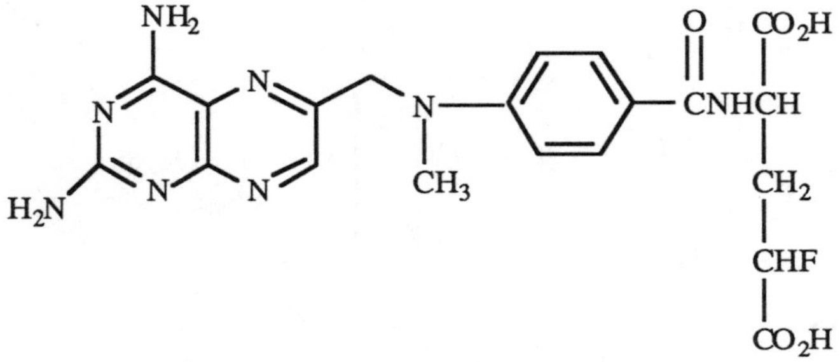

1 (FMTX)

FMTX is a potent inhibitor of the target enzyme for MTX, dihydrofolate reductase (DHFR, EC 1.5.1.3), and is readily taken up by mammalian cells, but cannot be converted to poly-γ-glutamates because of the chain-terminating properties of the FGlu substituent (*6*). Therefore, FMTX is ideally suited to study the role of polyglutamate biosynthesis on MTX cytotoxicity using cultured mammalian cells. Since previous work had shown that the highly anionic MTX polyglutamates efflux very slowly from cells (*7*), we postulated that a drug which could not form polyglutamates would not be well retained by cells and would be less cytotoxic.

The Role of MTX Polyglutamate Formation in Cytotoxicity. In order to address directly the question of the role of polyglutamate formation on MTX cytotoxicity, we investigated the effect of varying exposure time. The results show that if cells (H35 hepatoma) are exposed to FMTX or MTX continuously for 72 hours, there is only a slight difference in cytotoxicity between the two drugs. If, however, the cells are exposed to the drugs for shorter periods of time (2-24 hours), and the cells then allowed to grow in the absence of drug, there is a marked difference in cytotoxicity (*6*). This is due to the fact that the highly charged (anionic) MTX polyglutamates, formed extensively even after only two hours exposure, are unable to leave the cells. In contrast, FMTX forms only very small amounts of the polyglutamates and, as a result, readily leaves the cell and no cytotoxicity is observed. Since both FMTX and MTX are potent inhibitors of DHFR, continuous exposure of the cells to either drug gives similar cytotoxicity.

On the surface, it would appear that a drug which could not form polyglutamates and therefore not be well retained by cells would be an inappropriate candidate for an antitumor drug. However, in common with all cancer chemotherapeutic agents, the dose-limiting toxicity of MTX is associated with non-tumor tissue (e.g., liver, marrow, etc.) and the highly retentive polyglutamate forms of MTX may contribute significantly to that toxicity. If polyglutamate synthesis contributes more to normal tissue toxicity than to tumor cytotoxicity, our results suggest that an improved therapeutic index (tumor toxicity:host toxicity) might be achieved with FMTX *in vivo* using a sustained release technique. Preliminary results with tumor-bearing mice using a continuous infusion pump show that FMTX has a markedly improved therapeutic index when compared to MTX under identical dosage conditions (P. Sunkara and J. K. Coward, unpublished data).

Stereochemical Differences in FMTX Action. As noted above, cell culture studies on the non-radioactive *erythro,threo* mixture suggested that the fluorinated analog is not converted to polyglutamate forms (*6*). That conclusion has been substantiated rigorously with radioactive (^{14}C or ^{3}H), diastereomerically pure forms of FMTX (*8*). Although less than 10% of either diastereomer of FMTX was converted to polyglutamate forms, the *erythro* diastereomer, eFMTX, led to more intracellular polyglutamates than the *threo* disastereomer, tFMTX. In ligand displacement experiments, it was shown that eFMTX binds to DHFR with kinetics indistinguishable from MTX, whereas tFMTX binds slightly less tightly than MTX to DHFR (*9*). Finally, transport experiments have shown that eFMTX binds to the reduced folate transport protein and enters or leaves cells with kinetics very similar to that observed with MTX, whereas tFMTX is not transported as effectively (*8*). Although eFMTX appears to be the best mimic of MTX in terms of DHFR binding and transport kinetics, it is converted slightly to the polyglutamate forms. In contrast, tFMTX binds to DHFR and the transport protein slightly less efficiently than MTX but no polyglutamate forms of tFMTX are observed in cell culture experiments. Thus, these two diastereomers of FMTX may be differentially useful for studies on the biochemical pharmacology of MTX.

Properties of the Putative FMTX Glutamylation Product. In order to investigate the properties of the small amounts of glutamate addition product(s) which might be derived from FMTX under long-term cell culture conditions, we have synthesized one such product; 4-amino-10-methylpteroyl-4-fluoroglutamyl-γ-glutamate, AMPteFGlu-γ-Glu (**2**), shown below as a mixture of four diastereomers (*10*). We investigated the properties of this compound as a substrate for FPGS

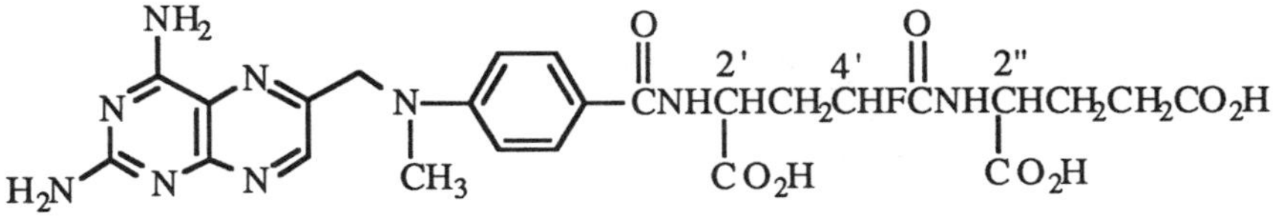

NH$_2$ ... N(CH$_3$)–C$_6$H$_4$–C(=O)NHCH(CO$_2$H)CH$_2$CHFC(=O)NHCH(CO$_2$H)CH$_2$CH$_2$CO$_2$H (positions 2', 4', 2")

2

	2'	4'	2"
AMPte (*threo*) FGlu-γ-Glu	S	S	S
	R	R	S
AMPte (*erythro*) FGlu-γ-Glu	S	R	S
	R	S	S

and γ-GH. The stereochemical heterogeneity of **2** precludes a detailed evaluation of the enzyme kinetics. However, it is clear from the data obtained thus far that **2** is a fair FPGS substrate with V_{max}/K_m = ca. 20% of the corresponding protio dipeptide, AMPteGlu-γ-Glu. We were surprised to find that **2** is an extremely poor γ-GH substrate when compared to the protio dipeptide over a concentration range of 0-300 μM. However, long-term incubation (48 hr.) of **2** with γ-GH leads to the expected product, FMTX. Since **2** is a reasonably potent inhibitor (I_{50}=10 μM) of the hydrolase, the poor substrate properties are due to a chemical reactivity problem and not to a problem in binding to the enzyme. One can speculate that the unexpected stability of **2** towards enzyme-catalyzed hydrolysis could be due to the effect of the fluoro substituent on the stability of the tetrahedral intermediate. Thus, breakdown of the intermediate could become rate-limiting in the case of **2**. Alternatively, ligation of the enzyme-bound Zn^{++} to the carbonyl oxygen of the scissile amide could be less facile due to the inductive effect of the α-fluoro substituent. However, in either case one would expect the α-fluoro acid product of the hydrolysis to be more readily ejected from the tetrahedral intermediate than the corresponding non-fluorinated acid. Elucidation of a detailed mechanism will require further investigation. Regardless of the mechanism, these results indicate that folate analogues such as **2** which contain fluoropeptides may accumulate in cells and be less susceptible to metabolic degradation via hydrolysis than the corresponding protio derivative.

Effects of Fluorine Substituents on Folate Biochemistry

Given the interesting effects of FGlu on MTX pharmacology described briefly above, we have synthesized the corresponding FGlu-containing analogue of folic acid, pteroyl(4-fluoro)glutamate (PteFGlu, **3**) and have obtained preliminary data on its biological properties (N.J. Licato, J.K. Coward, J.J. McGuire, and J. Galivan, unpublished data). The important role of folylpoly-γ-glutamates in carrying out

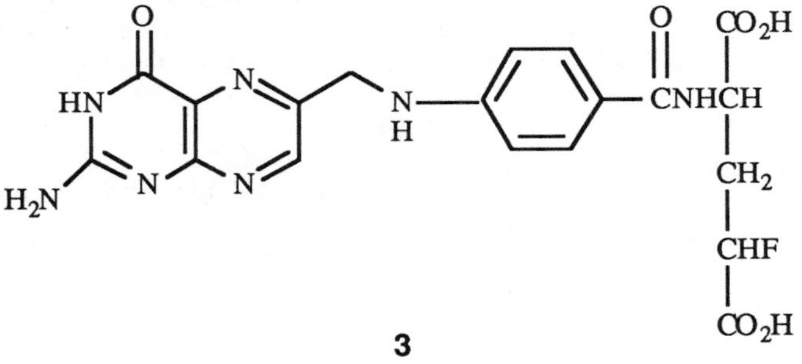

3

intracellular one-carbon biochemistry has been appreciated since the pioneering work of McBurney and Whitmore who showed that deletion of FPGS is a lethal mutation (*11*). Cells lacking the FPGS gene can grow only in the presence of all end products of folate-mediated metabolism. We have initiated complementary experiments using **3** as the chemical equivalent of deleting the FPGS gene.

As expected, the separated diastereomers of **3** are extremely poor FPGS substrates, with the *erythro* isomer again being slightly better than the *threo* isomer. This discrimination was noted previously for the diastereomers of FMTX. In addition, preliminary studies have shown that reduced derivatives of **3** are able to act as substrates for several folate-dependent enzymes; e.g., DHFR, thymidylate synthase, 10-formyltetrahydrofolate synthetase. With these data in hand, we have carried out cell culture studies to evaluate the ability of **3** to substitute for folic acid in the growth medium. Preliminary data show a dramatic effect of the single fluorine substitution and are completely consistent with the McBurney and Whitmore experiments (*11*). Thus, the concentration of **3** required to obtain 50% maximum cell growth (G_{50}) is 7 μM, regardless of the diastereomer used, whereas G_{50} for folic acid is 0.5 nM. This represents a difference of ca. 10,000-fold in the ability of the fluorinated folate analogue to support cell growth. We have not yet synthesized the fluorine-containing dipeptide derivative, PteFGlu-γ-Glu, in order to assess its sensitivity to hydrolysis catalyzed by γ-GH.

Thus far, the biological properties of **3** are in accord with the data described earlier in this paper for FMTX (**1**) and **2**. The overall effect of fluorine substitution on the biosynthetic (FPGS) reaction and on the hydrolytic (γ-GH) reaction is shown in Scheme II. The broad, open arrow focuses attention on the amide subject to hydrolysis in the reaction and its proximity to the single fluorine substitution.

Scheme II

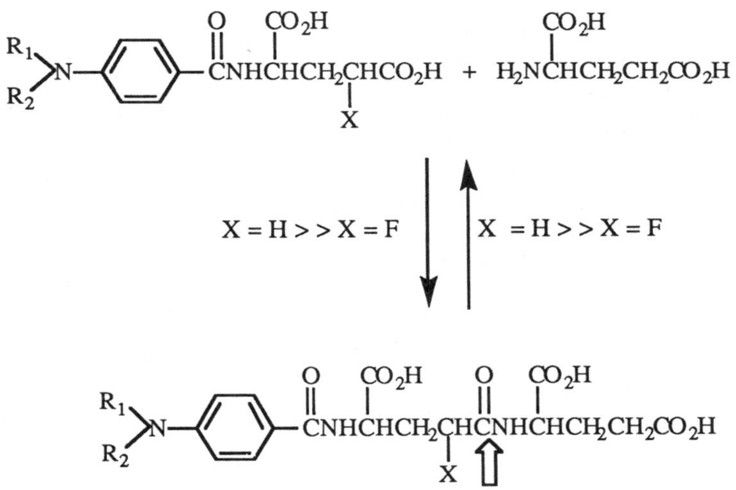

R_1R_2N–C_6H_4–$C(=O)NHCH(CO_2H)CH_2CH(X)CO_2H$ + $H_2NCH(CO_2H)CH_2CH_2CO_2H$

X = H >> X = F (forward) X = H >> X = F (reverse)

R_1R_2N–C_6H_4–$C(=O)NHCH(CO_2H)CH_2CH(X)C(=O)NHCH(CO_2H)CH_2CH_2CO_2H$ ⇧

Effects of 3,3-Difluoroglutamate on Folylpoly-γ-glutamate Synthetase

The presence of a second chiral center in FGlu has led to several interesting biochemical differences between the *erythro* and *threo* diastereomers, either as the free amino acid or when coupled to a pteroyl derivative, as in **1** and **3**. However, it should be appreciated from the results presented thus far that this stereochemical heterogeneity has been a problem in other cases, as in **2**. Even when it is possible to separate the diastereomeric blocked FGlu precursors, the resulting material is still racemic. Although we are currently working on a stereospecific synthesis of the four isomers of FGlu, we wished to study a fluorinated glutamate analogue which lacked the second chiral center of FGlu. The availability of 3'3-difluoroglutamic acid (F_2Glu) from the Merrell-Dow Research Institute allowed us to investigate the properties of this fluorine-containing amino acid in the FPGS-catalyzed reaction. Of particular interest was the fact that the measured pK_a=3.25 for the γ-COOH of F_2Glu (P. Bey, personal communication) is similar to that estimated for the γ-COOH of FGlu (*4*). Thus, if the chain terminating action of FGlu is due to the decreased pK_a of the γ-COOH (see introductory section above), F_2Glu should act in a similar fashion.

3,3-Difluoroglutamate: A Stimulator of Chain Elongation. We have recently completed an initial study of the ability of F_2Glu to act as an FPGS substrate (*12*). Much to our surprise, this new fluorine-containing glutamate analogue does not act as a chain-terminating substrate. On the contrary, F_2Glu is an extraordinarily effective substrate and actually facilitates the addition of several more F_2Glu or Glu residues to the growing poly-γ-(F_2)glutamate chain. With MTX (AMPteGlu) as the pteroyl substrate, F_2Glu is ligated to give the products of one and two additions; AMPteGlu-γ-F_2Glu and AMPteGlu-γ-F_2Glu-γ-F_2Glu, and very little unreacted MTX. In contrast, Glu is ligated much less effectively to give mainly unreacted MTX, some AMPteGlu-γ-Glu, and a trace of AMPteGlu-γ-Glu-γ-Glu. In addition, if one studies the kinetics of the addition reactions, it is clear that F_2Glu is much more rapidly incorporated than is Glu.

Chemical synthesis of the putative product(s) of the FPGS-catalyzed ligation of F_2Glu to MTX was precluded by the small amounts of F_2Glu available. Therefore, we have isolated the two products and have assessed their ability to act as substrates for two γ-GH's isolated from chick pancreas and hog kidney. These two hydrolases differentially cleave the poly-γ-glutamate chain at two distinct γ-glutamyl peptide bonds (Scheme III, open arrow (hog kidney) and filled arrow (chick pancreas)). The results of these enzyme-catalyzed hydrolyses provides strong evidence for the structures assigned to the reaction products as AMPteGlu-γ-F_2Glu and AMPteGlu-γ-F_2Glu-γ-F_2Glu (*12*). In addition, the rate of hydrolysis of the latter F_2Glu-containing peptide was shown to be considerably slower than the corresponding Glu-containing peptide. This result can be compared to the decreased rate of hydrolysis of **2** described above. In the case of **2**, the monofluoro substituent is located α to the scissile peptide bond, whereas in the case of AMPteGlu-γ-F_2Glu-γ-F_2Glu, the difluoro substituent is β to the scissile peptide bond.

It was of interest to establish the rate of glutamylation of an F_2Glu-containing product. Therefore, the dipeptide, AMPteGlu-γ-F_2Glu was used as a substrate for the FPGS-catalyzed ligation of glutamate. Consistent with what was observed in the product analysis described above, ligation of Glu to the F_2Glu-containing substrate was greatly enhanced relative to the ligation of the corresponding protio dipeptide, AMPteGlu-γ-Glu (*12*). The results of these experiments with F_2Glu are summarized, together with a summary of our results with FGlu, in Scheme III. Although unexpected, the chain elongation enhancing properties of F_2Glu may be exploitable in chemotherapy. Similarly, the unexpected resistance to enzyme-catalyzed hydrolysis

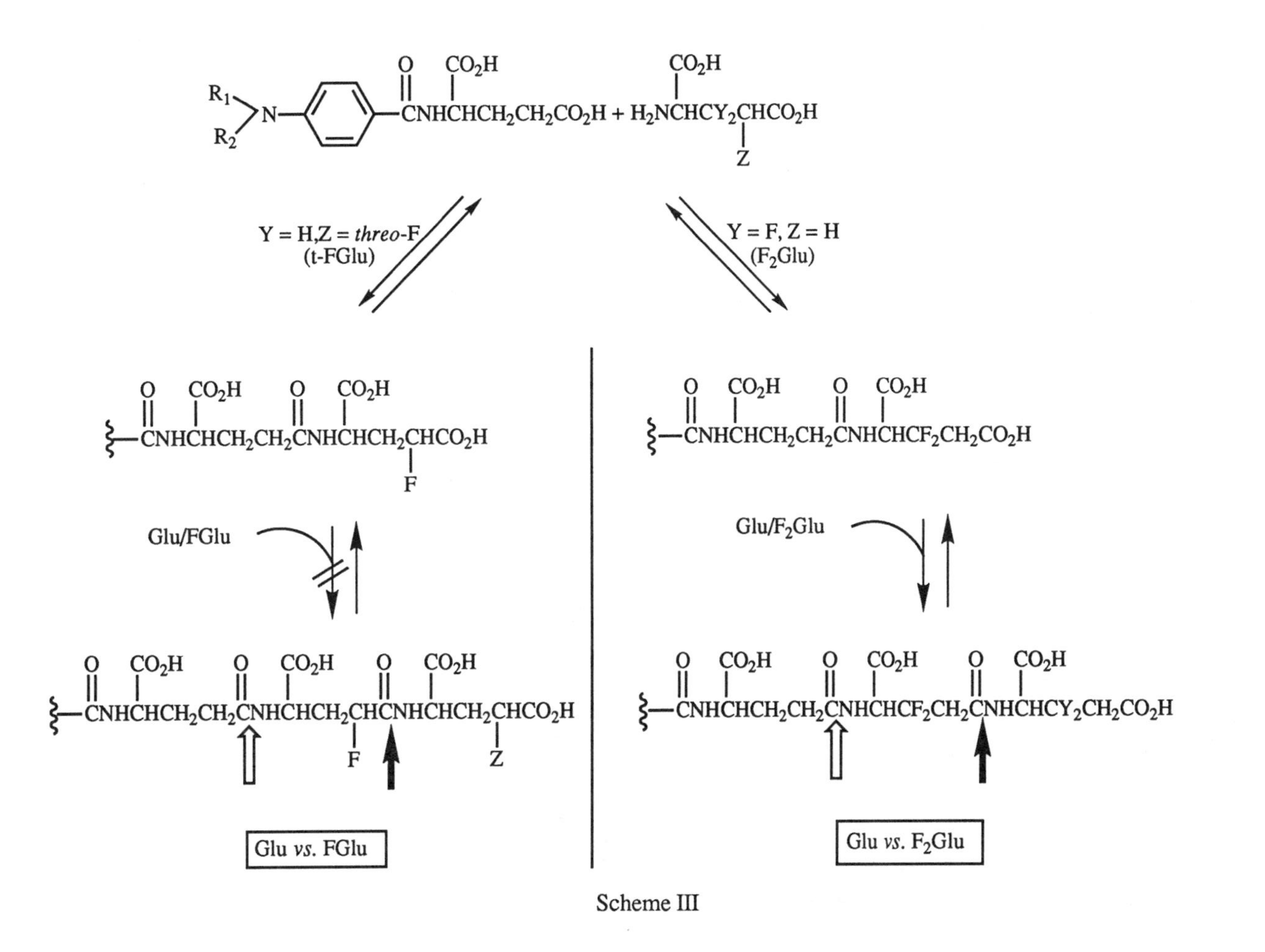
Y = H,Z = *threo*-F
(t-FGlu)
Y = F, Z = H
(F_2Glu)
Glu/FGlu
Glu/F_2Glu
Glu *vs.* FGlu
Glu *vs.* F_2Glu

Scheme III

by **2** and F_2Glu-containing peptides may provide access to more selective chemotherapeutic agents. Synthesis of compounds designed to test these hypotheses is underway in our laboratory at the present time.

Acknowledgments

This research was supported by grants from the National Cancer Institute, CA28097 (J.K.C.), CA25933 (J.G.), CA13038 (J.J.M.), and CA24538 (J.J.M.). We gratefully acknowledge Dr. Phillipe Bey of the Merrell-Dow Research Institute for his important contribution of supplying DL-3,3-difluoroglutamic acid for our research. We thank Jane MacDonald for preparation of the manuscript.

Literature Cited.

1. *Folates and Pterins*; Blakley, R.L.; Benkovic, S.J., Eds.; Wiley: New York, 1984; Vol 1.
2. McGuire, J.J.; Coward, J.K. In *Folates and Pterins*; Blakley, R.L.; Benkovic, S.J., Eds.; Wiley: New York, 1984; Vol 1; pp 136-190.
3. Matherly, L.H.; Seither, R.L.; Goldman, I.D. *Pharmacol. Ther.* **1987**, *35*, 27-56.
4. McGuire, J.J.; Coward, J.K. *J. Biol. Chem.* **1985**, *260*, 6747-6754.
5. Banerjee, R.V.; Shane, B.; McGuire, J.J.; Coward, J.K. *Biochemistry* **1988**, *27*, 9062-9070.
6. Galivan, J.; Inglese, J.; McGuire, J.J.; Nimec, Z; Coward, J.K. *Proc. Natl. Acad. Sci. U.S.A.* **1985**, *82*, 2598-2602.
7. Balinska, M.; Nimec, Z.; Galivan, J. *Arch. Biochem. Biophys.* **1982**, *216*, 466-476.
8. McGuire, J.J.; Graber, M.; Licato, N; Vincenz, C; Coward, J.K.; Nimec, Z; Galivan, J. *Cancer Res.* **1989**, *49*, 4517-4525.
9. McGuire, J.J.; Haile, W.H.; Coward, J.K. *Biochem. Pharmacol.* **1989**, *23*, 4321-4325.
10. Licato, N.J.; Coward, J.K.; Nimec, Z; Galivan, J.; Bolanowska, W; McGuire, J.J. *J. Med. Chem.* **1990**, *33*, 1022-1027.
11. McBurney, M.W.; Whitmore, G.F. *Cell* **1974**, *2*, 173-182.
12. McGuire, J.J.; Haile, W.H.; Bey, P.; Coward, J.K. *J. Biol. Chem.* **1990**, *265*, 14073-14079.

RECEIVED October 25, 1990

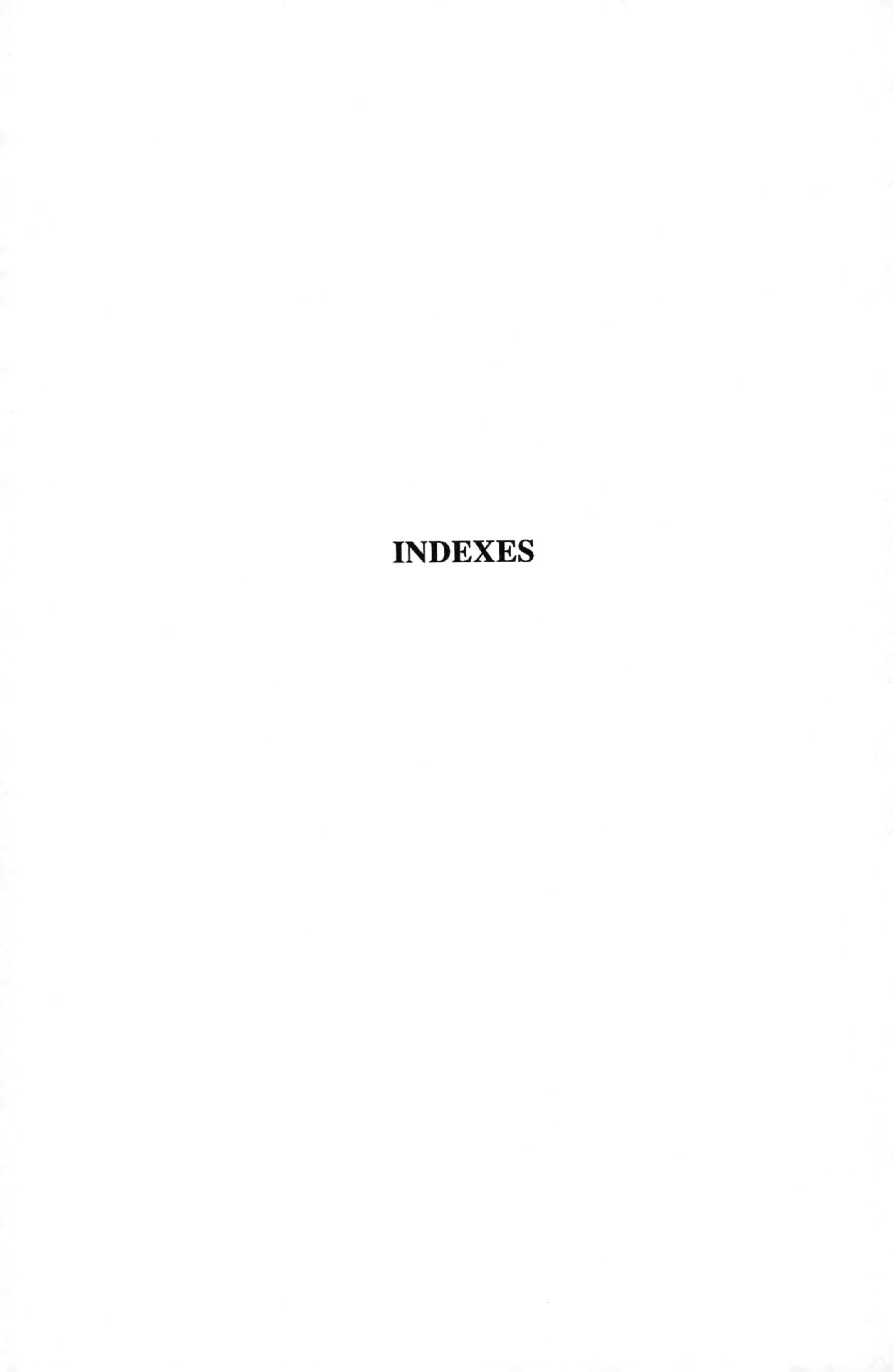

INDEXES

Author Index

Affiliation Index

Subject Index

A

B

C

G

H

I

K

L

M

N

O

P

Production: Kurt Schaub
Indexing: Deborah H. Steiner
Acquisition: A. Maureen R. Rouhi

Books printed and bound by Maple Press, York, PA

Paper meets minimum requirements of American National Standard for Information Sciences—Permanence of Paper for Printed Library Materials, ANSI Z39.48–1984 ∞